'Jack' Idriess was born in 1891 and served in the 5th Light Horse in the First World War. He returned to Australia to write *The Desert Column*, which was published following his huge success with *Prospecting for Gold*.

He went on to write 56 books and was largely responsible for popularising Australian writing at a time when local publishing was still not considered viable.

A small wiry mild-mannered man, Idriess was a wanderer and adventurer, with a vast pride in Australia, past, present and future.

ETT Imprint has published twenty-five books by Idriess including *Prospecting for Gold, Drums of Mer, Madman's Island, Flynn of the Inland, The Yellow Joss, Lasseter's Last Ride, Nemarluk, Gold-Dust and Ashes* and *The Red Chief*.

This illustrated 30th edition published by ETT Imprint, Exile Bay 2021

This book is copyright. Apart from any fair dealing for the purposes of private study, research, criticism or review, as permitted under the Copyright Act, no part may be reproduced by any process without written permission. Inquiries should be addressed to the publishers:

ETT IMPRINT
PO Box R1906
Royal Exchange NSW 1225 Australia

First published by Angus & Robertson Publishers Australia 1932. Reprinted 1932, (three times), 1933, 1934, 1935, 1936, 1937, 1939 (twice), 1941 (twice) 1944, 1951 (twice), 1965, 1973, 1982, 1985, 1986
Published by Cornstalk 1986
Published by CollinsAngusandRobertson 1993
First electronic edition published by ETT Imprint in 2017
Published by ETT Imprint in 2017, reprinted 2018, 2019, 2020, 2021

© Idriess Enterprises Pty Ltd, 2017, 2021

ISBN 978-1-922473-76-9 (paper)
ISBN 978-1-922473-77-6 (ebook)

Cover *Forgotten Heroes* by Geoff Harvey. Winner of the Gallipoli Art Prize 2021 (courtesy of the artist and the Gallipoli Club).
Cover and internal design by Tom Thompson
All photographs courtesy of the Australian War Memorial, as prepared for the 1951 illustrated edition.

THE DESERT COLUMN

Leaves from the Diary
of an Australian Trooper
in Gallipoli, Sinai
and Palestine during
World War One

Ion Idriess

Foreword by Harry Chauvel
Introduction By Ross Coulthart

ETT IMPRINT
Exile Bay

FOREWORD

I gladly send a few words of preface to Trooper Idriess' book on the Campaign in Sinai and Palestine. Not only is it a narrative of personal adventure which is full of interest, but it is, as far as I am aware, the only "soldier's" book yet written on that campaign. Several books have been written by officers and war-correspondents but in this the campaign is viewed entirely from the private soldier's point of view. It is of absorbing interest to a leader and should be to the general public.

At the same time there is an accuracy in the descriptions of operations which could only be provided by a singularly observant man. Idriess was, I think, above the average in this respect though I must say that the Australian Light Horseman was generally very quick in summing up a situation for himself. No doubt his early training in the wide spaces of the Australian bush had developed to an extraordinary degree his individuality, self-reliance and power of observation, and the particularly mobile style of fighting he was called upon to take part in suited him and brought out his special qualities far more than any trench warfare would have done.

In addition to giving a vivid description of the campaign as he saw it, Trooper Idriess also shows the interest that was taken in the Holy Land and its previous history. I think that this was not peculiar to the Australians but was common to all British troops, thanks very largely to the padres of all denominations who, intensely interested themselves, made it their business to interest others by lectures and personally conducted tours, etc.

I would commend this book to leaders who took part in the theatre of war with which it deals and also to the general public.

Harry Chauvel.
General Sir Harry Chauvel, G.C.M.G., K.C.B.
Commander Desert Mounted Corps

INTRODUCTION TO THE 100TH ANNIVERSARY EDITION

If Ion Idriess' stories were an American narrative, he would no doubt have inspired a dozen major feature movies. For over four decades he chased down the great yarns that have come to define Australia's sense of itself, pumping out over 50 books that told his countrymen and the rest of the world a prodigious array of stories that are now part of our national mythology.

Desert Column is one of his earliest books, a diary of his experiences during World War One, as a trooper in the 5th Light Horse Regiment that served first in Gallipoli and then in the Gaza desert. When Lt General Harry Chauvel, the Australian commander of the Desert Mounted Corps, made his desperate decision to order an attack on the town of Beersheba late in the afternoon of 31 October 1917, Trooper Idriess watched on as the Australians wrested victory from imminent catastrophe and seized the town.

It is impossible not to feel the rush of blood as you read Idriess' thrilling account of that extraordinary day. By this stage of the war any quaint 19th Century notions that modern machine-guns and artillery could be breached by cavalry had been dispelled by the awful realities of the horrors of the Western Front in France and Belgium. Perhaps that is why what happened that day on the outskirts of this Gaza desert town is remembered and immortalised to this day as one of the great feats of the war—precisely because it was such a preposterous decision that took the Turks by surprise.

The Australians and New Zealand soldiers and their horses of the mounted Anzac force were near collapse from lack of water and heat exhaustion and hours of British infantry attacks had failed to seize the town. As Idriess watched on, his mates from the 4th Light Horse began their charge on the town near sunset. His account of the 4th Light Horse's desperate final charge is breath-taking.

"Then someone shouted, pointing through the sunset … There, at the steady trot, was regiment after regiment, squadron after squadron coming, coming, coming! It was just half-light, they were distinct yet indistinct. The Turkish guns blazed at those hazy horsemen but they came steadily on."

Desert Column gives the reader a ringside seat to the realities of service as a mounted Anzac soldier, especially during the last great massed cavalry charge in recent modern history. It is irrepressibly laconic Australian in its prose and dry wit. What Ion Idriess made his career on was the notion that there were distinctively Australian stories that could replace the jingoistic and often nationalistically turgid myths of British Empire. As a writer, Idriess traced the origins of Australia's sense of itself.

Ross Coulthart

Trooper Idriess, 1914.

AUTHOR'S NOTES

The "Desert Column" is more than my diary. It is myself, I began the diary as we crowded the decks off Gallipoli and watched the first shells crash into Turkish soil. Gradually it grew to be a mania: I would whip out the little book and note, immediately, anything exciting that was happening. As the years dragged on, my haversack became full of little notebooks. These memories in tabloid form are my sole souvenirs of the War, except of course stray bits of shrapnel, bomb, and high explosive splinters which nearly every soldier collected.

The diary was a very young soldier's idea. He thought that if he survived shot and shell and sickness, he would like, when he came to be an old man, to be able to read exactly what his feelings were when "things were happening." Have a private picture show all his own, as it were, to refresh his memory.

Hence, all that has been written in this diary records my thoughts and feelings at that very moment.

Naturally they were many in nearly four years of active war and eventually necessitated the throwing away of my iron rations to find room in the haversack for the little notebooks. What the "Heads" would have said had they found out, goodness only knows.

Despite the fact that brevity had to be a watchword when I wrote, fully twenty thousand words have been cut from the diary in order that it may appear in book for at a reasonable price to the public.

I never thought the diary would appear in book form. But a proud sister, in whose care they were, forwarded the little notebooks to the publishers with the ultimatum that they must be published.

So, here marches again the Desert Column.

ION L. IDRIESS.
The Desert Column

1

TRANSPORT LUTZOW, CAPE HELLAS, DARDANELLES, MAY 18TH, 1915

Evening—An elusive vista of hills behind the dim outline of ships, a faint boom-oom-oom, then, like bursting stars, the shells struck a hillside here, there! over there! here again! everywhere!

Intense excitement amongst the 2nd Light Horse Brigade as our transport glided to anchorage. What a roar of voices as an extra vivid flame crashed on a black hill! Then night settled down and the bombardment ceased. At midnight a furious outburst of rifle and machine-gun fire broke forth from the land; but most of us slept, quite soundly.

Next morning—We awoke to the boom! boom! boomoomm-mmm of big guns. We scrambled up on deck to gaze at a medley of busy war- and supply-ships of intriguing kinds around us, all facing the low-lying hills that rose sheer from the sea. Among the foothills, like giant mushrooms, gleamed rows of tents with, in one distinct square, the orderly lines of horses. It was the English and French encampment. Came a harsh buzz overhead and 'planes soared across the brightening sky right over the enemy's position. They reminded me of hawks soaring above the farmyard as they gazed down seeking chickens. Bang! a fleecy smoke-puff, eerie in its startling appearance, burst far below a 'plane, to drift gently out into a perfect little cloud. Another cloud! higher up this time. Another! another! another! Each closer, closer, closer!

What an excited jabber of talk as we watch new clouds bursting faster and nearer the now rapidly climbing 'plane! Ah! see that puff high above the 'plane! There's another; look, another! They are all around it! And so time goes on and another 'plane drones along and cloud puffs follow her while the great birds circle and rise and soar and swoop, seeking the hidden batteries—and neither are hit.

And now, from far astern, forges a picture of titanic strength and energy—a French cruiser ploughing the foam. We drift closer inland and the boom-oom-oomm grows until we feel a shudder in it. We creep closer still and it becomes a roar, alive with a shaking sound—and the face of each man among us mirrors the excitement of his own queer personality.

Torpedo-boat destroyers, mine-layers, little craft all spiked with machine-guns and anti-aircraft guns buzz around us like hornets guarding a big fat prize. As the day wears on a lively bombardment shakes the shore and ricochets out to sea. One point of the land is fast enveloped by cloud-wreaths from exploding shells. What hell it must be for those alive there now!

…The guns have ceased. We are told to clean up the ship, as after we land she is to steam back to Alexandria loaded with wounded. It is a perfect morning, bright, breezy, and calm. It feels good to be alive and in such cheery company.

…About dinner-time a Turkish fort started a brisk cannonade upon one of the land ramps. We dropped our tucker knives and ran on deck just in time to see a shell explode on a battery galloping out. The head and tail-end of the battery-team galloped on through the smoke haze like maddened gnomes in a devil's pit, but the middle was ended. A cruiser raced close inshore, slewed broadside on and thundered in crashes of sound that rocked and echoed and smashed amongst the hills—and silenced that fort's guns. As seen from our transport the English camp is squeezed upon one small flat hill with the Turks fronting it and battering it from left and right.

…The afternoon we have spent in watching aeroplanes circling over the Turkish lines. One circled high over the ship, showing two glaring rings of red, white, and blue. In our inexperience we thought these were really impudent bull's-eyes. The Turks dotted the clear air with ineffectual shrapnel-bursts.

…Just before sundown a British four-funnelled cruiser opened fire only a few hundred yards from us. As the crash of her guns jumped through the Lutzow we answered with a rousing cheer, and instantly it seemed our rigging was crowded with men all excited and cheering as the big guns' talk, and roaring acclamations as the terrible shells exploded on the hidden Turkish position in columns of dust and smoke.

A big French cruiser has just ploughed past us full speed ahead—her mighty engines humming. Her band played us "Tipperary." How we cheered her! We are moving. We have just glided past the firing cruiser—our cheers were drowned in the explosive roar of her guns the whole great steel mechanism was a throbbing inferno of sound that fanned our faces. I can see our crowd going mad under excitement, like the 9th Battalion, when we go into action. "No smoking," and "Lights out" tonight is the order. We are steaming fourteen miles farther up the coast to the Australians' position—and what awaits us there.

And so the day has passed. While the guns have boomed some of us have watched, some sharpened bayonets, a few played cards, and some lay down below and joked and laughed.

Next morning—We steamed here with all lights out, moving stealthily through a black night, anchoring twice in fear of submarines. This is the Australasian camp, which has cost Australia and New Zealand so dearly. From the shore, all through the night after we arrived, came a ceaseless rattle of rifle-fire that swelled into roaring waves of sound made harsher by the burr-rrr-rrrr of machine-guns.

The intensity of the rifle-fire has ceased, but with the sun comes the deep boom! boom! oom-oommz from the warships. Opposite it is a tiny beach which rises abruptly into cliffs merging into steep peaks on a gloomy range of big, dark, scrub-covered hills. Mists curl rather drearily over the larger hills but at the beach the sun glints on stacks of ammunition cases, and dugouts and numerous queer things littered about. Cloud-puffs are continually forming over the beach—Shrapnel!

On either side of us are silhouetted the masts of battleships: the deep echoes of guns roll sullenly over the water. The "Old Brig." was on shore and has brought us the news of last night's fighting:

Two divisions of Turks, numbering fifteen thousand each, reinforced the Turkish position yesterday and last night attacked the trenches with the intention of driving the Australians into the sea.

The Australians and En Zeds waited until the Turkish charge was within fifty yards and then every man blazed away, the machine-guns especially, firing with deadly effect. This morning five thousand Turks are lying before the Australian trenches. There were a hundred and twenty-seven men and four officers of the First Light Horse Brigade killed. The infantry casualties we do not know.

The guns are boom-boom-booming. We are soon to land!

After breakfast—A big grey boat loaded with khaki men has steamed in from the outer sea. A fussy torpedo boat destroyer has just hurried the first crowd ashore. They are apparently Australians—a virile looking crowd, rather hard faced.

The shrapnel is bursting directly in front of that landing crowd now—from here it appears to be exploding above the first hill. Apparently the Turkish gunners cannot quite get the range. What ho—she bumps! A shell has crashed right into their boat! What a lovely time is awaiting us!

Our landing-party is ready. What oiling of rifles; excitement; laughing and swearing … Here come the destroyers, racing back hell for leather for more loads. Looks as if men are at a premium.

After dinner—The last boatload of men from the other ship has just raced shorewards. They are New Zealanders. As their packed vessels sped by we yelled from our crowded decks the old Cairo sayings: "Si-eda!

Talla-hena bint!" "Have you got a piastre?" and the pet sayings of the Tommies. They sounded comical with Australian voices imitating the English accents: "Has your mawther got a Ba-by?" "Have you been to Cairo?" etc.

…We are not to land until to-morrow. All are disgusted, but I suppose the "Heads" know best. The mists have long since evaporated. It is a beautiful day. We can see the shore distinctly, where our first battalions made Australian history.

What a seemingly impossible task they were set! The landing-place looks a sheer line of rocky cliffs, the abrupt hills frowning under their grey undergrowth. Cliffs and hills and gullies were swarming with Turks and machine guns at the Landing. It must have been a supreme bayonet charge, as awful as its success was miraculous.

And above survivors and reinforcements are now shrapnel-puffs, with much higher and to the right a scouting biplane like a droning bird. The cruisers farther to our right are roaring a devil's tattoo. Smoke drifts lazily across the water and wisps away. We are landing after all. The boat is already alongside.

…A Squadron's turn came. We tumbled down the ladders and packed tightly all over the tiny steamer's deck. Our kit felt massive, everything felt like haste, even the small steamer looked excited and puffy. Some men were rather quiet, but the majority laughed and joked. I wish I was good humoured. I swore when a big fellow tramped on my toe.

We moved off for the shore all ears to the pop, pop, pop of rifle-shots. Smoke fairly belched from the toy funnel: I suppose the sweating devils below were shoving the coal into her.

Nothing happened until we got closer inshore and the bushes on the hillsides began to take shape. Then whizzz, then ping, ping, ping, ping. By jove, rifle-bullets! Whizz-zz, smack! and a shrill receding whistle as the bullet ricocheted off the water.

We opened our eyes rather inquiringly, two or three laughed—then we all laughed. I know I felt a queer, excited warming at the stomach too. What if one of the damned things smacked into a man before he had a chance!

After that, we listened for the whizz-zz and laughed as the bullet would strike the funnel or smack off the water alongside. They were stray bullets perhaps, but they were straying uncomfortably close.

Our steamer pulled up with a grunt and rattle close inshore as two launches raced out to meet us. Men tumbled aboard and the packed launches raced for the beach, the busy captain roaring to us to get below because the shrapnel would be here any moment!

He had hardly closed his mouth when something came tearing in a shriek through the air and—bang! Bullets burst on the water, clattered on the sides of the boat, screeched through the funnel. How we tumbled down below! How we tried to pack into that little black hole—as if there were room!

There came another screaming whine bursting into a splitting bang! another shower of bullets and hissing of lead-sprayed water.

Someone laughed loudly. Lots of us laughed, some smiled, some shouted derisively advising the far-away Turks where to aim. But they sent seven well-aimed shells right around us and in that very short time the captain had up-anchored and we cleared right out for the horizon. Presently we crept in again, and anchored and went while the going was good. Were jolly glad to step on shore among huge stacks of ammunition and stores. Men were bathing—so strange it seemed that men were dying too. Men were toiling among the heavy stacks of stores, men trudged all over that tiny beach in ragged, clay-stained uniforms, their familiar Australian faces cheerful and grimy under sprouting beards.

Pock marking the first steep hillside were scores of dugouts, like rough black kennels. A few Indian soldiers were loading ammunition on mules. Zigzag tracks were cut along the big hillside. Close inshore were overturned boats, all shell-smashed, relics of the Landing. We glanced at the sinister humpy of sandbags that is the dressing-shed. A long line of wounded were lying on the ground waiting their turn. The doctors looked busy; they appeared awfully workmanlike. One of our fools had to make a joke of course as we hurried past. He called to us that he could see one of the doctors and an orderly cutting off a man's arm.

We climbed up the steep hill path, joking with the toil stained warriors who were cooking their evening meal, or toiling at the ammunition boxes, or lying like tired brown men about their tiny dugouts. Then we filed through a trench that led to the back of the hill and came out in a gloomy, narrow valley all tortuous and fissured as it wound through a sort of basin at the bottom of the big, sombre hills. We now faced many hills gutted with gorges, overshadowed a mile farther ahead by a flat rampart of cliffy peaks. An occasional shrapnel-shell screamed overhead. The whizz, zip, zip, zip of bullets became definite and unfriendly.

And so we climbed the back of the big hill that faces the sea. We are digging holes to get in out of the way of the shrapnel. Quite close one of our hidden Australian guns replies to the Turks and makes a monstrous row. The steep hills are covered with a dense, prickly shrub.

…We have just been called to arms. I suppose we are going into the trenches.

2

SHRAPNEL GULLY, GALLIPOLI.

May—The regiment has been all night under fire. Rifle-fire started suddenly. In one minute we could not hear ourselves speak. Came a hoarse whisper: "Fill magazines!" We jammed in the clips. My fingers tingled as the empty clips grated out of the magazine. Another horse whisper, and we fled down the shadowed ravines that gouge our big hill.

Then ugh-ugh bang! Shrapnel burst above us in an instantaneous black-grey cloud of smoke: bushes around bent as if under a hail-storm. We scrambled on a little faster, instinctively ducking our heads from the storm. Again came that long-drawn scream to apparently split with its own velocity—bang!—and hail whipped the bushes around. We crouched low as we slithered down the ravines, grasping any handy bush, steadying ourselves with our rifle-butts, slipping and sometimes falling to scramble sheepishly up to the instant joke. The sun was sinking: it was a creepy feeling among those black hills: we did not seem to know what was happening: we were hurrying somewhere to kill men and be killed. Our own battery answered the Turks. The rifle-fire grew to a roar that drowned the voice of the man beside me. I felt as a stone-age man might feel if volcanoes all around him suddenly spat fire and roared. Doctor Dods in front suddenly fell on his knees. I caught my breath—it would be awful if the doctor were the first man killed! But he got up again and scrambled on. He did this several times. Instinctively I understood. The doctor had been to the South African war! The next time he ducked, I ducked. When he scrambled up, I followed suit—so did others, a sort of automatic ducking all along the line. Thus we quickly learnt the best, the quickest chance of dodging shrapnel. We had to. Immediately that scream came tearing directly overhead we would duck down flat. The doctor glanced around and we laughed. He is a long man and very soldierly, but has no dignity at all when he flops down so. He screwed up his face and winked over his shoulder. Then we all rushed forward again and burst through the bushes that lined a little road meandering through the hills. We were in the little valley again, with the big black hills enveloping us. They call it Shrapnel Gully.

It got coldly dark. If a man were home he'd be just at the sliprails letting go his horse before he went in to tea. Or a city man would be pushing open the garden gate. Bang! bang! whizz whizz, zip, zip, zip, zip, bang! zip, zipp, bang! bang! bang!—Hell's orchestra, with additions, as the shrapnel burst in vicious balls of flame within the hollow basin: on the sides of the hills: above the hills—everywhere! We dashed up that narrow "Road," five hundred of us. Whizz, whizz, bang! then rrr-rrr-rr-rrr in a rattling stutter, tut—tut—tut tuttuttuttuttuttut, zip, zip, ping, zip, smack, zip! Bullets, machine-gun and rifle, whined down from over the big hill before us, not all aimed directly at us, seemingly a chronic shower flying over the tops of the infantry trenches to rain down the gully. Harry Begourie stood absolutely still in front of me, his hand to his head, the blood hurrying down his right cheek. The doctor led him away. We ran on. "Halt!"

We all crouched by the roadside, among the bushes, by something solid, or in a sheltering hole. A man near me sighed in the darkness as he found a shallow dugout. For an hour we lived there, clinging to cold mother earth, an invisible regiment whispering among the bushes; a rustle when a man cautiously lifted himself to shift his bayonet more comfortably from under his belly.

Overhead screamed the big shells from our own guns, travelling towards the Turks and they were shrilled in chorus by the Turkish shells as they criss-crossed down into us. The shells made hell's row in the dark and when exploding close dazzled our eyes with sheets of flame. Jagged fragments screamed into the bushes or struck rocks and screeched piercingly away. My body was alertly passive, but the mind was curiously thinking, "So this is War!" The rifle-bullets in the bushes were the devil. With the night they seemed to lose a lot of their nasty threatening force. They just chirped among the bushes like busy canaries. One landed under my nose with a gentle squeak, but the flying gravel stung my lips viciously. One tore the heel clean off Burns's boot; the force swung his leg around. He laughed and called out to the invisible quartermaster for a "new issue!" Those unseen canaries were real nervy, one might thud into a man's back at any second!

Muffled footsteps slow and heavy came down the dark gully. We strained our eyes as the shadowy forms of Stretcher-bearers went by. I shivered involuntarily to the vanishing pad of their feet, for a man's hand hung limply over a stretcher.

Then "Forward!" We arose and stumbled on up the gully. Not far though. We dived for the bushes again, for anywhere, and stayed there the whole night, cramped and shivering and cold, listening to the big shells trying to burst the very air above, to the ominous roar of rifle fire just ahead, to those damned canaries among the bushes. The rifle-roar in front made us certain that the Turks were attacking only a few yards away. We knew nothing. Most of us did not know where our officers were. They were scattered amongst us of course, but I doubt if they knew where they were.

"We are to be used as reserves for the trenches!" someone bawled in my ear.

"How the hell do you know?" I yelled, but got no reply.

When a sort of shadowy moonlight came there sounded a cracking rumble coming down the gully. We held our breath. "Guns!" Around a bend in the gully came some sort of animal dragging a squeaking, funny looking cart, a little cart that rocked and swayed over the uneven gully-bed We could not see distinctly, but two other carts creaked slowly behind. The drivers walked, cloaked and silent, creaking past us with never a word. A voice from the night said distinctly, "Don't their legs wobble funny!" The dead men's legs were hanging over the rear of the carts.

At daylight we crawled out of our retreats, the officers trying to sort out the hopeless muddle of squadrons and troops. Gillespie sat on a stone, six feet of patience, trying to get some dirt out of his eye.

Anyway, we got back to our "camp." Now we have shifted and are digging out fresh lodgings as yesterday's are too abundantly supplied with shrapnel. Only a very few of us were hit last night. Providence may have smiled on us, but I'll bet our quick instinct for crawling into any available "burrow" shielded us a lot, too. The cracking of rifles is still menacing, but mainly from snipers. The rifles never cease; every minute they crack! crack! crack!

Strange! a few birds are flying about, merrily chirping, while the air is trilling with death!

I was a fool this morning. I saw such a bonza pair of Turkish top-boots, almost brand new, sticking up out of the ground near my new dugout. I pulled at one boot and out came part of a Turk's leg.

All along the paths leading to and from the trenches are the graves of the Poor Aussies who have been shot. A man grows wary where he walks—many of the graves are so shallow. They could literally call this place "Death's Gully." I've only been here a few hours, but, by Jove, I've seen some dead men. And old dead ones make their presence felt right up and down this great gully.

May 22nd—A party of us volunteered for a sapping job last night. We left camp at eleven and followed the road, which is the gully bottom, meandering up to the firing-line. Across the gully are built sandbag barricades which shield a man just a little from the death-traps along the road. We would bend our heads and run to a big barricade, lean against the bags until we panted back our breath, then dive around the corner and rush for the next barricade. The bullets that flew in between each barricade did not lend wings to our feet for nothing could have made us run faster. A few hundred yards ahead of us and high up is the firing-line, perched precariously on a circle of frowning cliffs. The Turks have an especial trench up there which commands our "road." This trench is filled with expert snipers, unerring shots who have killed God only knows how many of our men when coming along the road.

None of our party were hit. Eventually we reached the farthest bunch of sandbags, stacked higgledy-piggledy on a shadowy mound directly beneath the big cliffs by Quinn's Post. It was pitch dark up by the cliffs. On the cliff and hill peaks the rifles fired like spitting needles of flame. The firing was not heavy, but numerous bullets came thrillingly close.

Our object was to cut a trench from a sap, through the little rise back towards the Gully, and thus save the necessity of walking along that particular danger-spot of the tragic road.

The night, away from the bases of the hills, was only semi-dark. The bullets coming so far were mostly high shots flying over the tops of our trenches which clung to the cliffy hills above.

We set to work, a shadowy line of us working in pairs and, by Jove, we did work. We made that dirt fly digging ourselves in. I had a queer impression of grave-diggers in the night who were digging their own graves.

The bullets began coming faster and faster and each man felt mighty glad when he had scooped a hole down three feet and thus had partial shelter. We could dig on our hands and knees now. Then one poor chap, Watson of H Squadron, was shot clean through the head.

We worked in a sweating hurry as the night's inferno began, thankful that the tall hills sheltered us greatly from artillery-fire. But there came a rain of vicious bullets. One smacked a clod of earth down my neck and the old spine shivered to the grisly dirt. Heaven only knew what germs were in that soil from the battlefields of ages. We belted into the digging again. A bullet sighed clean through Nix's hat. It rained bullets for a while and we crouched in our holes like anxious mice, wishing they were deeper. To an awful roar and a flame that crimsoned the sky we gazed awe-struck above our burrows. We breathed intense relief when we realized it was only the Japanese bombs, and not the whole world blowing up. We had brought those bombs along in the *Lutzow*—a present, I believe, from the Japanese government. Our fellows were now passing the gift on to the Turks and I did not envy the poor devils the receiving. At the same time, the knowledge that we Australians can fight courageously in these bloody battles, using and struggling against the most terrible modern weapons, filled me with a deep feeling of satisfaction.

Straight up from the cliff's black edge there rocketed skywards two flaring bombs. They descended directly into the Turkish trenches. For breathless seconds the rifle-fire ceased, then came a tearing roar that shook the very ground. In streaks of blackened flame there spewed up a smoke-cloud blacker than the night. Some Turks were panic-stricken. In inky silhouette we glimpsed them, like toy men, away up on the cliffs as they sprang from their trenches to run. But the air vibrated to the machine-gun and rifle-fire that was turned on them. And thus the game went on all through the night. After our working shift, we lost no time hurrying down that "zip—zipping" road to "camp."

…Last night we were sapping again, right up at Quinn's Post. None of my particular crowd were shot, but we are so tired and sleepy! The air was foetid with the smell of dead men.

We are frightfully tired, but have to shift camp once more.

…The infantry are quite cut up—not over their terrible losses, but because of one man, Simpson Kirkpatrick I think his name is. He was known everywhere as "Murph. and his Donk." At the Landing he commandeered a donkey and ever since has been coming and going from the distant firing-line to the beach with wounded men. He worked day and night, plodding along unscathed under fire till all thought he must be protected by supernatural means. His colonel long ago told him to carry on all on his own; to do whatever he liked and go wherever he liked. He has been a little army of mercy all on his own. Yesterday morning, I think it was, he went up the valley and stopped by the Water Guard where he generally had breakfast. It wasn't ready so he went on, calling, "Never mind, give me a good dinner when I come back."

He never came back. Coming along the valley holding two wounded men to the donkey he was shot through the heart. Both wounded men were wounded again.

3

The warships have been circling the bay with the torpedo-boat destroyers racing around them. A Turkish submarine has sneaked in, and to see the way that insignificant little wasp is stirring up these big grey ships would be amusing—if one did not realize the significance.

Our new camp is on a rugged hill near the sea. It is hard for a man to understand the various positions and zones of fighting, but it appears to me that the little bay is the hub of everything. Practically from the landing place Shrapnel Gully meanders up between the hills to its cliffy head at Quinn's Post, about a mile from the beach. Our fighting-line seems a crescent with the Post the farthermost middle, from whence both flanks bend back to the coast. We are within the crescent and the Turks, with all the country behind them, are pressing us to drive us into the sea.

The line of the crescent might be five miles. We are in continual anticipation of the Turks' breaking through for they hold the commanding positions in overwhelming numbers. I can see hellish fighting if they do smash through our trenches. They will find battalions and regiments scattered amongst these hills, and if all of us have to fight in our own little groups, the slaughter amongst the Turks will be more terrible than the punishment we will have to take. This "anticipation" of berserk things liable to happen at any moment is an unexpected phase of warfare for me. But then, I am just a learner.

We went for a swim this evening. As the Turks sent an occasional shrapnel screaming across the wee beach, our bathing was in running dips. The beach is strewn with discarded equipment, broken rifles, numerous mess-tins, and water-bottles, all with shrapnel holes through them, lengths of barbed wire, jagged stakes, torn haversacks, and now and again a trampled-on felt hat with the little hole and black blood patch. These are the relics of the landing of the first battalions and each tragic lot of flotsam tells its own story. I picked up a sand-dirtied photo of a woman and kiddy, with a bullet hole through it. The graves of the men line the beach, their shrapnel-torn boats lie overturned at the water's edge. It seems pitiful waste, men and everything smashed—and hearts in Australia too.

The Turks have been given a seven hours' armistice to bury their dead. They are rotting in thousands in front of our trenches. The air is dreadful. It is raining.

By Caesar! we have just had ten shrapnel-shells dropped right into camp. We were all sitting by our dugouts talking and laughing. Immediately the first shell burst there was not a man to be seen. The hill reminded me of a huge rabbit burrow when a man with a gun comes along. As each shell burst, a rousing roar of laughter went up. Not a man was hurt. It seems a real miracle: shrapnel is so deadly. By Jove, here they come again. Good night. We can hear the shrapnel-bullets pelting the ground around our dugouts.

May 25th—We have just witnessed the torpedoing of the Triumph. While having dinner we saw a lively commotion among the craft in the bay. Then small boats raced out from the shore. We dropped our mess-tins and rushed to high ground and there, close inshore, with a destroyer standing by and firing like blazes, was the battleship already lying on her side. Busy boats were taking the Jack Tars off while from all over the bay arose the dense smoke of big ships running for their lives. Slowly the Triumph moved over—we could see the sailors leaping down—she moved over quite stricken. Green water seemed to climb across her deck—we could see down her funnels—then she turned completely over and floated bottom upwards.

Water-spouts of steam belched up from beside her. Then the Turks tore the shrapnel into us and we ran for our burrows. The Turks had wonderful targets as we massed on the hills and gazed down, but they did not fire until she sank. I don't know whether they were "sporty" or whether they were too excited gazing at their own triumph.

The battleship has completely sunk.

The Turkish batteries are throwing us our evening bye-bye of shrapnel. We hear that over a hundred men went down with the Triumph. One shrapnel-shell has just got five men of the 2nd Light Horse. The stretcher-bearers are busy day and night. It is sad: the lads are such fine chaps it is hard to see them die.

…It is raining.

May 26th—Rifle-fire has been very quiet last night and to-day, but the Turks have got a lot of our men with shrapnel. The cruisers have all cleared out. We miss their protecting shells. The Turks are seizing the opportunity to pump more shells into us and the reply from our few guns is feeble by comparison. Our boats sank a submarine last night. Three torpedoes were fired at the Vengeance, but luckily all missed. We hear that the submarines are German and Austrian. Torpedo-boat destroyers are landing more troops on the beach. The Turks' shrapnel is exploding right over the crowded vessels. The destroyers have raced farther out.

We are getting better tucker now and have had a free issue of tobacco, which is appreciated more than words can say.

...One of the Turkish shells has just struck a destroyer loaded with troops. Luckily it only killed two and wounded ten. The troops are landing—jumping from the boats in a wading plunge to the beach. The shrapnel is exploding above them—their backs are bent as they run across the beach seeking shelter.

The snipers shot seventeen of our men to-day in one spot alone in the gully.

May 27th—The Majestic was torpedoed last night. It is stupefying: those massive ships stricken so suddenly. It is woe to us, for the help of their great guns feels almost like human support.

...We are being shelled with shrapnel again; the damn things are screaming overhead and bursting with frightful crashes. Hardly a man in the 5th that has not experienced some miraculous escapes. Steaming hot fragments of shell have plunged into our dugouts by day and by night, bullets have pierced men's hats and equipment, some have nicked the puttees of men as they slept. And yet we have only had a few men hit.

...We are expecting a momentous move soon.

...Our big howitzer is replying to the enemy's fire. She invariably does when the firing gets too hot for us. Of course, we are only one little group. There are other battalions and regiments for miles. They all have troubles of their own. So our howitzer looks after us and it is warmly cheering to hear that big shell tearing through the air overhead on its vengeful errand, and then the distant bang!—fair on the enemy's trenches, we hope, or better still, on some hidden gun.

...We were bathing just now when a shell came and wounded McDonald and Liddell, both of my own troop. Bathing is off—until to-morrow.

May 28th—Snipers shot fifteen Aussies this morning and shrapnel got four of our regimental A.M.C. men. It is heartbreaking to see so many men killed and maimed when they are not in the actual firing-line. No matter on what peaceful errand we go, death goes too. We never know whether we will wake up alive.

...The enemy has kept surprisingly quiet these last few days and nights

...I was down on the beach just now on fatigue duty. A man had his leg blown off. The doctors were working at it—it looked like a big red lump of beef. War is a sickening thing.

They are bombarding us with shrapnel. Their aim is getting startlingly close. They must have got better observation-posts during the armistice, or shifted their guns nearer. Did I mention they buried three thousand Jacko dead and many of our own during the armistice?

…Last report is that the Turks' most vicious eighteen pounder is silenced. We sincerely hope so.

…To-night is the first night that we have had no shrapnel. We are sitting by our dugouts on the hillside, little groups singing, others smoking, others lousing themselves. The colonel's gramophone is playing the "Marseillaise," and the rest of us are silently watching a beautiful, peaceful scene. The sun is sinking, a golden ball, behind the island of Imbros. Over a sea of deepest blue destroyers are quietly gliding: the hospital-ships are anchored closer inshore, near numbers of smaller craft which seldom dare to anchor; behind us the hills are darkening: the Aegean is peaceful and one star is in the sky. As I write, one of our own guns has broken the peace; a bomb has burst away up in the trenches and now comes the vicious crackle of rifle-fire. Soon the first fury of the night will storm upon us.

May 29th—It is a cold, grey dawn. Death is bursting all around us now, but plenty of the boys are crawling out and lighting their little fires. A long, tanned bushman is kneeling down, blowing carefully into flame a few dry sticks; a smoke-cloud from an exploded shrapnel-shell is drifting into the air above him. A counter-jumper on the gully opposite is bemoaning the swine who pinched his scanty supply of wood, and invites the unknown lousy thief to come out and fight. And one fair-haired chap is combing his nice wavy hair with a dirty old broken comb.

…They are calling out for stretcher-bearers now. I don't know how many are hurt. The Indian stretcher bearers are busy too… What miraculous escapes! A shell has just burst in front of me; ten feet below two infantry chaps are cooking their breakfast. They were splattered with smoke and earth but refuse to leave their cooking. As I write, another shell has burst beneath them, but still the obstinate goats won't budge. Another has come and this time they grabbed their pots and ran. We all laughed. By Caesar! It is about time they did run. A fourth shell has come and where their fireplace was is now a cloud of smoke, ashes, earth, and fragments of shell. Shells are bursting amongst us all over the hill and up and down the tiny gully. A rumour is flying around that the Turks broke into our trenches last night. They never got out again alive, I'll bet. Johnny is sending his big stuff over this morning—high explosive mixed with the shrapnel-shells. A shrapnel bursts in the air and sprays the ground with hundreds and thousands of bullets, according to the size of the shell. One gun alone, firing quickly, can make a regiment feel as if it were under the rifle-fire of five hundred men. But when batteries of guns get going—! High explosive has its own terror. The shell itself plonks fair into the ground then explodes with a fearful crash broadcasting jagged fragments of shell.

...Finished breakfast under their cursed shrapnel-fire. They are "searching" for the Indian battery now, but their creeping shells are exploding above us—unseen marksmen sending unseen death into unseen men. These artillery duels mean sheer hell for us chaps in between. Not half a stone's throw away from us, just across the gully are the dugouts of an infantry battalion. Owing to the direction that the shells are coming these poor chaps are getting it far worse than we are, for the shrapnel bullets are pelting right into the mouths of their dugouts. One shell has exploded fair in a dugout and blown the four men up into the air.

Another two are now hit. If it were not for our dugouts we could not live. They are perfect shelters against shrapnel only, except when a shell explodes fair in or above them, or just at the angle where the bullets will strike down into the dugout. The dugout is no protection though against a direct hit by a high explosive. It is getting fearfully hot now, shells are exploding every few seconds; the row is an inferno made hellish by the hot smell of fumes. The cry is becoming continuous, "Stretcher-bearers!" "Stretcher-bearers!" "Doctor!" as more and more of our poor chaps get hit. Good luck to all the medical men! We have the shelter of our dugouts, but they run out into this hail of shrapnel directly the cry goes up. The pity of it is that we cannot fire a shot in return.

Bits of earth come crumbling down from my dugout walls as I write, lying on my back. It is hard work writing that way, especially with the blasted ground rocking underneath. One poor chap opposite has just had hard luck. His mate was wounded in the leg. He knelt up, undoing his mate's puttees. A fragment of shell whizzed into the dugout and took the top of his head clean off.

...This horrible fire is easing off. I stopped writing, just lay flat out and shivered as those great shells sizzled, seemingly just over my nose. We have had five heart-choking hours of it. It seems longer.

4

May 30th—It does not seem possible! A few hundred yards distant from this hurtling bombardment comes the command: "Slope Arm-ms!" "Order Arm-ms!" etc. Drilling men on a battlefield! Surely those in command have not gone mad?

…There was bitter fighting at Quinn's Post last night. Seventy stretcher-bearers are carrying our dead and wounded down from the trenches … Just now we heard a riotous hullabaloo, and jumping from our dugouts saw a hare racing across a sandy hillock. All hands cheered the hare; they yelled and laughed. It must have thought us queer folk … Things are quiet now … A man was just shot dead in front of me. He was a little infantry lad, quite a boy, with snowy hair that looked comical above his clean white singlet. I was going for water. He stepped out of a dugout and walked down the path ahead, whistling. I was puffing the old pipe, while carrying a dozen water-bottles. Just as we were crossing Shrapnel Gully he suddenly flung up his water-bottles, wheeled around, and stared for one startled second, even as he crumpled to my feet. In seconds his hair was scarlet, his clean white singlet all crimson.

…Last night the Turks exploded a mine under Quinn's Post. I thought the whole dashed hill had blown up. We sprang from our dugouts with bayonets drawn, expecting God knows what to come howling down upon us. But it was only one of those "local" affairs. It is all damned local. It depends just on what particular "local" spot a man happens to be. But what a lovely sensation, to go to sleep and wake up on top of an exploding mine!

A lot of poor chaps up at the Post were killed. The Turks threw hundreds of bombs into the shattered trench; distant though we were we could hear their howling "Allahs!" as they charged in the roarings of the explosions. The survivors of the Post were driven back into the support trenches to the fanatical delight of the Turks. But our supports and the survivors united and went over the top with bomb and bayonet and mad Australian strength and cut the Turks to pieces in the very frenzy of their victory.

…We hear now that the Turkish loss in last night's little "local" fight was fifteen hundred. Out of the portion of the trench occupied by one company alone of the 15th Battalion they have just dragged fifty-nine bayoneted Turks. Between the Post and the first Turkish trenches, the enemy dead lie in huddled heaps.

...These little destroyers amuse me. One particularly cheeky spitfire sneaks close inshore to suddenly whip around and blaze away at the Turkish batteries like a fiery terrier barking at a big dog hiding behind a fence. The noise of her guns is bigger than herself.

...Fifty of our men were killed at Quinn's Post last night, including Captain Quinn. But the heaps of Turkish dead lying between the trenches proved their sacrifice to Allah to be in vain.

...Firing is very subdued today. During breakfast we watched two big 'planes being shelled by the Turks. As usual they could not hit them.

Sunday afternoon—Our guns have started a lively bombardment. The Turkish reply sounds condescending. The rifles are cracking from the trenches now. I wonder if an attack is anticipated. Rumour has it that the English and French have captured the big hill Achi Baba, which bombards us so heavily. But it is only a rumour ... The rifle-fire is running the crescent circle of trenches for miles. Sounds like an attack. The bombardment is furious now; screeching devils burst above us every few seconds. Columns of smoke and earth are flying skyward as if the land were vomiting under the high explosive. Smoke-clouds are drifting over the warships down on the bay. The thunder of their broadsides crashes against the Shore and is blasted back over the intensely blue sea. And yet with all this whining death about us there are actually a few men out of their dugouts tending their cooking-fires. No wonder the Turks call us the "mad bushmens."

...A mine has exploded up in the trenches, its echoes are still rolling over the hills. Brigades of machine-guns are stuttering as if coughing their hearts out. The cooking enthusiasts have ducked. Two shells burst immediately above them. My appetite has gone... The Turk's sharp-pointed bullets have a peculiarly piercing sound, especially when they land only a few feet away. They seem to split up then and hiss away to hell ... A man lying flat in his dugout under heavy shell-fire feels so pathetically helpless; the deep, coughing grunts of the shells as they crash down from the heavens turn his belly to water, it does mine, anyway. When a blasted shell comes screaming in that tearing way they have when they are going to burst very close, I feel I have no belly at all and in imagination draw my knees up to my chin, in nerve tingling anticipation. But I don't move at all, just lie perfectly still and "freeze," waiting.

...I have just had a narrow escape: a shrapnel-bullet has grazed my right knee. Only a scratch ... The firing is easing down.

May 31st—Last night my troop and A troop were called out for trench-digging. My knee was too stiff for me to go. If my troop goes into the trenches without me it will be just too miserable: I want to be in the first fight the 5th are in. I don't care so very much afterwards.

…Six of our crowd were shot yesterday. It is damned hard our getting picked off like this and not being able to fire a shot in return … Last night they buried a few of our fellows and a lot of Turks who were killed yesterday afternoon … There has been a big explosion out at sea. The destroyers are racing towards it, their sides awash with foam. I hope no more of our ships have been torpedoed … We are detailed for the trenches this afternoon, and I am sick.

…Occasional shells have been coming and going throughout the day. Corporal "Noisy" was badly wounded yesterday. Poor Noisy, he was the regimental comedian. The doctor is making me stay behind for a day or two. As the regiment is only going sapping after all, I do not mind. I hope my blooming knee has not got this septic poisoning the doctor is always warning us about.

June 1st—I have just watched a poor devil of an infantryman being carried down on a stretcher. Half his face was shot away and he was trying to sing "Tipperary." And yet, here I am lying in my dugout, in no pain, fretting like a great kid because I could not march off with the regiment. My blooming knee is now poisoned. The 7th Light Horse took over our dugouts during the early hours of the morning. I hobbled to the 7th doctor this morning and when I returned someone had pinched my poor supply of jam. The world seemed a dinkum hard place. I complained to the man in the next dugout that they might have left a sick man's jam alone. He has shared his dinner with me. He has two onions; I have a tin of beef. He is going to fry them to-night and we are going halves. He is going to keep my water-bottle filled. If he gets shot I think I will nearly howl. I cannot boil tea—can't even crawl now to find the wood—and steel biscuits with water and salt tinned beef is no cure for a poisoned leg. Things are very quiet. Only a few stray shells passed today … The Turks are just beginning to shrapnel us again.

…We have just witnessed a shameful, horrible thing. Part of the infantry next to us consistently drill their men on a tiny plot of flat ground fronting their dugouts, even when the shells are falling only a few hundred yards, and less, away, and when they know that we are right under the glasses of the Turkish artillery observation posts. Such tragic idiocy, drilling soldiers on a battlefield that is night and day under fire! It is what we call a "Gawk Act." A shell has burst right amongst the thickly packed lines and six of the poor fellows are hit, two down, two being carried, and two limping for their dugouts. Others are hit who can get away. Another shell has burst while they were carrying one poor wretch in. The little patch of "Parade Ground" was torn up by the bullets, just as a dusty road by a fury of hailstones. Another shell has burst above the ground but they are all away now. One poor fellow is crying out in agony. It was a case of sheer murder. The officers responsible should be shot like mad dogs.

…I have had such a splendid tea. Tinned beef mixed with onions and fried in bacon fat; biscuits and cheese; biscuits and jam, and hot tea. Even some mustard which my good Samaritan pinched from a transport. If only I had a match to light the pipe! My rugged friend is short of matches himself and I have not the heart to hint that I would like one.

…I wish my leg would get better. I can't even hobble about now. It's a bit miserable lying here of nights. No sleep, and all my mates away. The stars twinkle ever so high up.

June 2nd—Tragedy visited us last night. The roof of a large dugout fell in and smothered three men. They are holding the burial service now. It is something unusual. The infantry are standing about in quiet groups, listening. I think these three are the first of all the poor fellows buried here to the strains of "Nearer My God to Thee."

…I felt it would have to come. I am down at the beach at the Ambulance Hospital and this afternoon have to go off to the hospital-ship. The ambulance men say thirteen of the infantry were struck by that shrapnel yesterday, four being killed, It was sheer murder on the part of the officers responsible!

…A great crowd of us are on the minesweeper now. One man, whose body and arm are a mass of bandages is affording much amusement to all by trying to eat his slice of bread and jam with only a third of his mouth visible; the hole is hardly big enough to blow away the flies.

…I feel strangely sick and feverish. My troop leader, Mr McLaughlin, is sick aboard too. We have been swapping tales of misery. It is amusing in a way; if only there was not so much pain about. The ambulance men on the boat are kind and gentle: one is making me a bed on the deck now. There is such a crowd of wounded everywhere.

…I am getting a bit of ease at last, thank heaven! There's one thing strikingly noticeable about this shipload of misery—I suppose every man growls, but almost every one growls only to himself. My safety valve is in this diary. It keeps my mind surprisingly occupied. And besides, if I really live through this war. I want to read through the old diary in after years, and remember what war was really like—as I saw it, anyway.

…The officers on this boat cannot do enough for us. Sometimes I feel miserable watching them trying hard to help the poor maimed chaps, and they have got nothing to help them with. They wear partly khaki and partly naval togs. Why are wounded men dumped on a naval boat? I suppose the two big hospital-ships are full up.

Light Horsemen waiting to hit the beaches at Gallipoli.
Headquarters of the 5th and 6th Light Horse at Rosenthal Ridge, Anzac.

5

LEMNOS ISLAND,
S. S. FRANCONIA.

June 5th—My little woes overwhelmed me. Dropped the diary with my tail picked it up again to kill time.

The A.M.C. men coming on night-shift washed me. You could not possibly imagine how I felt towards those men. Five days lying in a dugout without a wash, and sick.

Had a wretched night and only smiled when the bundle of bandages beside me turned his pain-dimmed eyes next morning and whispered: "I hope I did not keep you awake last night."

My own moaning had troubled my conscience, too.

Breakfast came in, a bowl of hot porridge, a slice of bread and jam and a cup of tea with milk in it. Such a welcome meal! though we could eat so little.

Then our vessel steamed up against a towering ship. Iron doors clanged open in her side and into that black cavern they carried the badly hurt men. We others followed somehow. Someone shouted: "Go downstairs and have breakfast." So we found ourselves huddling down a nice wide stairway emerging into a regal room where there were rows and rows of long tables and cushioned chairs. And such a quaint medley of voices, mostly Scotch and English mess orderlies.

Porridge, bread and butter, jam and coffee were put before us. We tried to eat but were too sick. Hard lines! Presently a few drifted away, but most of us lay down to rest just where we were: right in the way but there seemed to be nowhere to go and many of the chaps were very ill. My own leg became unbearable, so I dragged myself back up the stairs. No one seemed to know where the doctors were. A procession of bandaged Cripples were dragging themselves up another lot of stairs. I followed, and we shuffled and crawled into a huge room. There were two doctors and a few A.M.C. orderlies. The doctors' faces were heavy from want of sleep, their eyes had a starey look. They worked quietly and continuously and I saw at once that if there is an etiquette in a surgery then it was right out of place here. Instruments were picked up one after another and quickly used; if a man had to feel pain then he had to feel it; there was no time for any niceties that might alleviate pain in minor operations. There were so many waiting. All the boys took it as silently as they possibly could. They were stretched out in grotesque lines, all converging towards the doctors. As a man was treated, he would fall aside somehow and his line would wriggle, crawl, shuffle, or hop one up, then lie quietly until it was time to crawl up for the next man's turn.

One doctor was a big, kind-hearted Frenchman. I could see that saturated with the misery around him as he must have been, still he did not like hurting the men. At long last my turn came and I quickly and thankfully found out that the French doctor knew his business. But it made me sick mentally to find out how bad my knee really was. It must have been rotting right down the leg. I crawled downstairs again and lay in absolute misery on the cold cement floor until long after the last bugle-call had blown. One man told me there were three thousand beds in this ship but you had to get a long thin steward in a blue uniform to get you one. Another said a sergeant was in charge.

No one of us know anything about it, of course. It is just simply a huge ship, pitiably under-staffed with doctors, crowded over and over again with sick and wounded men. Away in the ship somewhere are wards where the doctors are working day and night with the cot-cases, not curing them because they have no time, but just trying to keep life in as many as possible until we get to a hospital. Up our end of the ship are the men who can look after themselves.

I called out to a Red Cross sergeant major; he told me to stop the long thin steward if I should see him pass. I think it a crime that a man wearing his badges should see a feverish man lying on a dirty deck and pass him by. I am growling, of course, but then all men growl when they are sick, and I am only growling to my diary anyway. At last, a big English Tommy with a bandage over one eye lifted me on his shoulder and carried me downstairs. He put me on this little bunk with its straw mattress, and tucked me around with this blue blanket. What a relief! For the first time in days my leg has stopped throbbing, and how much easier it is lying in bunk … The first bugle-call for tea woke me from a half-sleep, with the burning feeling gone. Some good Samaritan let me lean on his shoulder while I climbed those weary stairs to the tea saloon. It hurts horribly to move about.

…One of my bunk companions lit a cigarette for me. He told me that the ladies of Athens had sent the wounded soldiers thousands of cigarettes. They are done up in very pretty little packets of twelve cigarettes each. If only those ladies of Athens could know how their gift has been appreciated by all these hurt men!

…Last night I only woke three times. Such a splendid rest after four nights' sleeplessness. It hurt horribly to stumble to breakfast. Is it getting better or worse?

...Was only a little bit feverish to-day, and the knee does not hurt when I'm lying down, but it is awful when I've got to walk ... Had a bad two hours before dinner. Got feverish and couldn't walk. Yet a persistent idea of mine is that without decent tucker this leg won't get right. I called out to a long-legged, homeless-looking Australian. He soon came hobbling back with a plate of beef and vegetables, a big pannikin of broth and an appetizing grin. You can just imagine what that hot soup tasted like. But I could only peck at the meat and vegetables, after all.

...Feel real good this afternoon and am going to tackle those stairs again for tea. But, my heavens, the bugs are awful.

Next day—I'm continuing this tale of woe. Why shouldn't I! Nobody loves me! Anyway, if a man does get through this war, he'll have something to give him a fit of the blues just by reading up these notes and remembering things. But I know this growl is justified. Here it is. From ten in the morning until five in the evening, all men (except the distant cot-cases) are supposed to be on deck. At ten o'clock a ship's officer examines the ship, with a great flourish of trumpets. What ridiculous nonsense it is. Here are hundreds of men, not supposed to be seriously wounded, many of them limping about as I am myself, the majority, between them, hurt in almost every part of the human body; and yet they are debarred from the only thing this under staffed, overcrowded ship can give—rest. I got over the difficulty. On the mornings I can manage it I crawl up the stairs and get my leg dressed, then crawl back again. When the Tommy M.P.S. comes down to clear the ship for inspection I tell the sergeant to carry me up on deck if he wants to. He is nonplussed. You see, there are hundreds of us. Some just stare at him with fever-glazed eyes.

Why not have a good growl while I am about it? It is so wearisome lying here. Now, the Tommy doctor and the French one are working hard all day long dressing the wounds of those hundreds of men. My doctor has four assistants who undo the bandages, but three of them appear quite incapable of doing up a simple bandage. It is a shame the things that happen here daily. This might explain some of the things I am trying to tell. An assistant took the bandage off my leg and then started to pick hairs and fluff from the inflamed wound with a squat thumb-nail under which the dirt was thick.

The doctor can't watch all and dress our wounds at the same time. I have, seen him suddenly turn around and "go" for an assistant in a most fierce manner. But with all us men in the big room he is working at the very fever-pitch of mental and physical strength. And all of us, especially those whose wounds are paining, are quite willing to let anything be done to them if only the doctor will dress their wounds and give them ease. The poisoned pus accumulates and hurts like hell. We have got a very kindly feeling towards the big French doctor. The men who patronize the Tommy doctor, like him too, Both doctors are as gentle as they can be, but they have such a terrible lot of men to get through.

…I feel a little better this morning, so much so that I'm going to growl again. The lice are accursed things. I've broken out in a red rash all over the body from their bites. And it's hell lying here feverish with the bugs biting a man, I suppose they go and bite some other poor devil and fill him with poisoned blood.

By Jove. writing these notes when I'm able passes away the time and helps a man.

There goes that accursed raucous-toned bugle for the first dinner sitting. And I can hobble up there to-day. It will be quite an adventure. This is the first day I've not been feverish for some days past.

…Here is another howl. We only had two small slices of bread and jam and half a pannikin of tea for the evening meal. I've been whispered that we can buy buns at 1d. a piece, and coffee at 2d, a pannikin, and so fill up that way. It seems to me like kicking a roan when he's down. Of course, I feel hungrier than I suppose I really am. If the blooming old leg would get better I wouldn't care if they gave me bully-beef and biscuits, even if this is a hospital-ship.

…I wonder what this ship really is classed as. The lucky few who can hobble about say that she has no green band around her, and no red cross. They say she has been waiting to sail for Alexandria for these last twelve days, unload her wounded, then sail to England for troops. But it has been far too risky to sail. I suppose she will make a great dash for the sea one of these fine nights and if she gets through we will be very proud of the fact of having saved some green and white paint. But if she is torpedoed, what a cry there will be of a hospital ship with two thousand helpless men aboard having been sent to the bottom, "murderously" torpedoed, etc.

…This is a huge ship, but I am unable to look over it and can't describe it. It is strangely quiet and subdued, very different to a troopship. With this number aboard a troopship she would be a hive of Babel many times multiplied. Our little squadron in their dining hut at Mahdi kicked up fifteen times more row than comes from all this great ship's dining-halls.

…Mr McLaughlin came down and saw me this morning. I'm glad he is better, but he looks jolly miserable on it … I'm blest if the Tommy sergeant hasn't been trying to urge me up on deck to get into a life-belt. It appears we are actually off to-night. The men are being shown their posts should anything happen. I pray we may truly leave for Alexandria, but I wouldn't climb those stairs unnecessarily for all the life-belts in the world.

…Hurrah! I believe it is really, really true. If only we do go to Alexandria, my knee will have a chance. I have had grave doubts lately. All our wounds are only dressed. Nothing else can be done to them. A man would inevitably have to lose his leg.

…Thank God! We are off right enough, amidst frantic cheering. There are glad hearts aboard to-night, which may seem a strange thing to say of this ship of misery. But you see, many others besides me realize that our only hope is a hospital and attention. May we have a safe voyage, above all a swift one. I wish I could be on deck now while we are going out; all is excitement; they say there are many ships anchored around us.

Our great sniper, Billy Sing (left) with his spotter, at Gallipoli.
(Officially credited with causing over 200 enemy casualties.)

6

EGYPTIAN GOVERNMENT HOSPITAL, ALEXANDRIA.

June 28th—Here I am, in Alexandria, growling as usual, tired of having been in bed so long. The old leg, under efficient treatment, healed rapidly, but it seems to have broken out again. I employed the time lying here by thinking out a couple of war inventions. I forwarded the plans to the Brigadier and to Admiral Robinson. Naturally I got no reply. However, it helped time pass. I saw Trembath in the Deaconess Hospital yesterday. He is still the same cheerful old growler.

…Such a lot of our fellows have been wiped out. Poor Allan Williams died as he had lived, a Christian and a gentleman … Am in Mustapha Base now. In a few days numbers of us patched-up chaps are going back.

…We have just heard of the death of Colonel Harris. The 3rd Infantry Brigade were attacking the trenches in front of Tasmania Post, while the 2nd Light Horse Brigade masked the enemy's fire. The colonel was shot through the neck and died within two minutes: the 5th will be sorry. The colonel was a disciplinarian, but not oppressively so. We all respected him, and many liked him.

August 23rd—At sea again, in *A 25 Huntsend*, which looks very much like the *Lutzow* repainted. The Heads have got numbers of us on guard at absurd places. One man is guarding a dry water-tap. I am on guard at a tap too, but there is about six inches of rusty water in the tank, so I'll have to be careful. I hope the Turks don't fly aboard at night and pinch it. Most of the guard on duty last night did not know what they were guarding. The narrow place where I am on duty now is so stuffily hot that it is an effort to breathe. If the officer on duty came down and caught me writing in the diary, the skipper would hang me to the yardarm, or make me walk the plank. What a fuss I'd kick up, though, if they confiscated the diary, which they certainly would. Why don't the authorities leave the men alone? They want all the rest they can get now … This is a cosmopolitan shipload. Officers and men and non-coms, infantry, light horse, artillery, A.M.C. men. Australians, En Zeds and a few Tommies.

…There was a most peculiar incident just now. Right behind the ship, as if sneaking after us, there appeared tiny billows reminiscent of a submerged craft. We stared in an almost breathless anticipation, but our vessel gained speed until now those ominous waves are barely visible. We thought it was the wake of the steamer at first, but we have circled like a terrified deer and can still see that curious phenomenon.

…A transport was torpedoed in these waters several days ago. It is curious to notice how the chaps are taking the chances: the majority apparently don't care a tinker's cuss. But they are much quieter than they were on their maiden trip; and I hear odd groups surmising between pipe-puffs as to their chances of getting back this time. We all know what to expect ahead of us. It was the old Australian spirit leaving Cairo and Alexandria yesterday. Yelling and cheering, laughing and joking at the least little thing. That is the spirit that will never die.

…It is a beautifully calm morning. There are stowaways on board, same as last trip. Most are from the Ammunition Column and after twelve months' soldiering want to see fighting. They'll soon see it!

August 25th — Anchored at Lemnos. The island is of low hills; destitute of vegetation, with small rocky shores running right into the sea. The harbour, apparently small, is packed with a puzzling fleet of ships. Everywhere are masts — big masts, small masts, short masts, long masts, and the fighting masts of ships of war. On the brown land in places are white lines of tents

August 26th — How well old Germany has helped us in this war. Alongside us is a huge vessel from which we are taking in water. It has the German eagle nailed to the funnel, while our own transport was a crack German steamer. Quite a number of other German vessels are visible. Kaiser Bill's navy slipped on its guarding of their merchantmen.

August 27th — The minesweeper is standing by. We are re-embarking for the final four hours' trip. There was a storm at sea last night. This morning driving rain and mist temporarily blot out the shipping. The rough sea looks cold. We hope it. is not raining on the Peninsula for the cold there is intense. We old hands are just taking things as they come. Whatever happens, must happen. That is the best way, after all. We have all our equipment ready. The pack of the Light Horseman on foot is heavy and clumsy. A man will do some slipping about if those Gallipoli hills are muddy.

Late afternoon — We have just re-embarked, walking across deck after deck of closely packed steamers to this other boat. As usual, the Australians were rowdy. The En Zeds howled their hair-raising Maori war song. Among all the hubbub and slowness of the re-embarkation the Australians were imitating the bleatings of a mob of sheep being yarded, made more realistic by the barkings of men who, of a certainty, had worked in many a station muster.

It was all very new and unusual to the numerous ships' and English officers, who looked on from all those crowded vessels. Some of them gazed in a sort of bewilderment at the apparently unruly crowd surging over the ships, but most of them enjoyed it immensely. The Tommies marched steadily beside us and their sedateness and dignity no doubt emphasized the comical aspect of the business … The sun is blazing cheerfully. I hope the sea outside is not over rough. We have no room to sit down, unless on another man's body.

August 28th—We sneaked into the Landing at ten o'clock last night. A hospital-ship was beautifully lighted on the still waters. Here and there fires gleamed on the old dark hills, and far down on the new Landing at Salt Lake. Those grim black hills waiting there seemed to spread a sinister atmosphere all over the bay. Desultory firing was going on with an odd sharp outburst of machine-gun fire at Salt Lake. All night long we were disembarking, only a few hundred men too, crouched shivering in those big iron barges. The steam pinnace took hours to tow the three barges the stone's throw to the shore.

We landed at daybreak, and we score of the 5th started out to try and find the old regiment, or I should say, the remnant. I met old Gus Gaunt coming over the hills. He looked a living skeleton, just yellow skin stretched tight over bones. He told me of our troop and all the old mates who are gone. Poor Fitzhannam, shot through the head yesterday morning. We plugged on up a goat track winding around the hills. The regiment was away on the right, at Chatham's Post.

…I have quite a cosy dugout in a trench, where I can lie and gaze down on the sea and see the cruisers stealing around the shores. I am detailed for the firing-line now.

Late afternoon—This trench sniping is intriguing. I am a crack shot, and was put on sniping. Not seeing any Turks I blazed away persistently at one of their loopholes. The invitation was presently answered. "Ping, ping, ping," Johnny Turk replied. One of his bullets flattened on the loophole plate, which suggests the wisdom of letting a sleeping dog lie. When he quietens down again I'll worry him some more. Perhaps one of us may get the other. The sandbags along the trench are all spattered with bloodstains.

August 29th—Put in a bad night, standing gazing through the dark towards the Turkish trenches; the blooming old leg was giving me beans before daylight. I could just see the top of the Turkish trenches in the half moonlight and used to snipe now and then. Some Turk answered snappily. I believe he must have aimed his rifle by daylight and set it in a vice, for his bullets repeatedly struck the loophole plate. It is a nasty sound when a bullet flattens on iron within an inch of a man's nose.

...I "spotted" awhile for Billy Sing this morning. Billy and I came down on the same boat from Townsville. He is a little chap, very dark, with a jet-black moustache and a goatee beard. A picturesque looking man killer. He is the crack sniper of the Anzacs. His tiny possy is perched in a commanding position high up in the trench. He does nothing but sniping. He has already shot one hundred and five Turks. He has a splendid telescope and through it I peered across at a distant loophole, just in time to see a Turkish face framed behind the loophole. He disappeared. A few minutes later, and part of his face appeared. That vanished. Five minutes later he cautiously gazed from a side angle through the loophole. I could see his moustache, his eyebrows, and part of his forehead. He disappeared. Then he showed all his face and disappeared. He didn't reappear again, though I kept turning the telescope back to his possy. At last, farther along the line, I spotted a man's face framed enquiringly in a loophole. He stayed there. Billy fired. The Turk vanished instantly, but with the telescope I could partly see the motion of men inside the trench picking him up. So it was one more man to Billy's tally.

...Dr Dods has just been hit in the shoulder by shrapnel, when he was out in the open attending a wounded man, an 11th Light Horse chap who had only been on the Peninsula a few hours. The news simply flew around Chatham's Post. Since the beginning of the regiment we have admired and respected Dr Dods. We are all relieved that he is not hit too badly.

August 30th—Last night was very quiet, just desultory firing rippling away down the line into silence. Our trench runs downhill with the barbed wire into the sea. There's a little destroyer that seems to have adopted the regiment, as a terrier does a man. Last night it sneaked in again, blazed away hell and fury at the Turkish trenches just opposite us, then whipped around and faded into the night as silently as she had come.

...A taube is buzzing overhead now. He visits us every day and drops a few bombs, but doesn't seem to do much harm. It was misery fighting to keep awake last night—had to rock to and fro and then fire a few shots and then rock again and force open my eyes with my fingers, or I would have fallen asleep in spite of everything. Part of the trench caved in last night. More work.

...Civilities were exchanged up at Lone Pine last night. Then, presently, the Turks (game men) sneaked right up to the trenches and slung in half a dozen bombs. They were "duds." Instead of relighting them and throwing them back as our boys generally do, the bombs were allowed to lie there until morning, and then examined. The fuses were instantaneous. If our chaps had put a match to them they would have instantly been blown to pieces. So Johnny Turk was "had."

Afternoon—I've just been indulging in a duel with a Turk, shot for shot. I'd fire, and the dust would fly up against his loophole. Then slowly and cautiously the tiny circle of light on the trench parapet which was Johnny's loophole would fill up with half his square, grim face. Watching like a cat watching a distant mouse-hole I'd see his rifle-muzzle slowly poke through the loophole, then a spurt of smoke with the crack—ping! and his bullet would plonk into the sandbag above my loophole. Then my turn. I'd wait with my rifle-sights levelled evenly at that distant tell-tale gleam of light, then immediately it was blotted out by his cautious face, I'd fire. Instantly he would duck. And vice versa, and so on. It was thrilling. I waited for each of my turns with every sense keyed to concert pitch, thrilled through and through. No doubt Johnny the Turk felt the same. I tried to kill him, and he tried to kill me. Yet we have never seen one another and never will.

…Our regimental trench system is called Chatham's Post … I wish there was not so much night work—the sleeplessness is cruel … A few evenings ago, some of the 6th Light Horse off duty were amusing themselves by playing Two-up behind the trenches. The Turks started shelling them, but the Two-up enthusiasts took no notice until presently a shrapnel bowled over three of them. So they picked up their wounded and retired casually into their dugouts, one of the wounded men arguing volubly that he had won the last toss.

7

August 31st—On outpost duty last night. The Turks were quiet. Our lively mascot far down on the black sea flashed a searchlight on to the Twin Trenches and Balkan Gun pits, blazed away with her snappy guns, then vanished into the night ... Outpost work is a bit nervy, waiting for Turks to arise up out of the dark and jab you with bayonets. My blooming leg ached last night ... Just had a delicious wash, the first for four days. I wish there was time to wash clothes, the lice are awful. I'm sick of cracking the rows of dirty white eggs along the seams of the trousers. Getting the water for our wash was humorous. To see Gus Gaunt running naked from the beach, trying not to spill a bucket of water while Turkish bullets hastened his feet, was really moving—likewise his language.

Evening—There has been desultory firing all day. Bursts of machine-gun bullets, odd shells coming with a roar, the usual bullets pinging and zipping around. The guns at present sound exactly like rumbling thunder. Johnny the Turk got two bullets fair into my loophole this morning. His shooting is improving The taube is buzzing overhead.

September 1st—Last night the machine-gun and rifle fire roared as it rolled away down towards the left flank. Our little destroyer was exceptionally cocky.

All was quite dark when suddenly the Balkan Gun Pits opposite were illuminated with ghastly clearness. From our shadowed trenches we distinctly saw the spurts of dust as our bullets struck the Turkish sandbags. A sharp *bang! bang! bang! bang!* slapped across the water. The Balkan Pits were struck by bursting stars of flame which flung sandbags, dust, and cloudy smoke into the air to drift away through the dazzling light. Then instant darkness, bringing a vengeful volley of rifle-fire with hundreds of bullets pinging into our parapet. But our heads were well below ground by then. It must be galling to Johnny Turk to have his parapets blown to dust while we joyously blaze at him and he dare not show up to fire back in return. Then, when all is "quiet" he has to work feverishly throughout the night to rebuild his parapet. And last night before dawn the destroyer came again and blew his new parapet to smithereens. His trench is now in ruins, and he daren't rise to fire a shot.

...Another "stunt" to-night is rumoured. A "stunt" means a raid. A line of men with bombs dangling from their belts and armed otherwise with any lethal weapon they fancy, creep over the parapet and snake their way down the hill-slope, followed by more men with fixed bayonets.

They creep ever closer towards the Turkish trenches and if they do not run into an enemy patrol or meet a sudden volley of machine-gun fire they get right to the parapet and hurl in their bombs. Following the explosives they and the bayonet men, with mad yells, jump down into the fume-filled trench and kill every Turk they can. They then rush back hell for leather to our own trench.

Our crowd have raided before, generally losing a few men. But now Johnny is awake to the joke and has all sorts of jokes of his own awaiting the raiding parties. So any man due for a raid, cannot help wondering whether he will see another dawn.

The "Old Bird" is a holy terror in these raids. He's only an exceptionally small chap and no youth either, but he is about the most murderous old devil in the regiment. He leads these raids with a hell of a yell as he jumps down into the trench, blazing to right and left with a sawn-off shotgun. An ordinary service revolver is no good to him.

The big Heads confiscated his little toy though; he blew a Turk's head clean off his shoulders and that wound put the show away. The Turkish Heads complained. It appears that through some international law or other a sawn-off shotgun is not allowed as a "weapon of war." The major is bitterly peeved.

…Heavy gun-fire comes rolling over the sea from Cape Belles. To-day is simply routine. A shower of shrapnel for breakfast. Then Johnny Turk behind his loophole, we behind ours, shot for shot. Johnny got one of the new 11th Light Horsemen last night.

…Sometimes Johnny is daring and dangerously cunning. Out of our loopholes and through artillery telescopes two of us are now watching five little pieces of dried-tip bush away out in No Man's Land. Telescopes are deadly aids. Behind that, oh so natural-looking, little bush is a "fox" hole which Johnny has dug in the night and painstakingly carried the tell-tale earth away. There are quite a number of these inconspicuous holes, artfully hidden by the bushes that cover the ground in front of his trench. Lying snug in the hole we are now watching is Johnny, his rifle poked through his bush. Thus, when he does fire, the momentary puff of smoke will be hidden. Ah, but we have patience—and our powerful telescopes show us even the wee black hole of his rifle-muzzle. I am putting the diary down now, for I am going to let the light in through my loophole and slowly pass a piece of bag behind it. Then, when Johnny fires at the supposed face, my mate—

…We have got Johnny.

September 2nd—Had a rest in the support trenches last night. It was great. Loused all my clothes. These woolly socks with a lot of fluff on them are favoured by the lice for laying eggs. But they like the seams of the trousers best. Warm, I suppose. A man is only a blooming incubator for them ... the stunt for last night was declared off ... In a charge, the 1st Light Horse lost a lot of men. Their wounded lay in front of the trenches for days and could not be brought in. Two wounded men in particular were lying dose to the Turkish parapet. Five of their mates went out to get them but were shot. One man managed to crawl back after two nights and a day, and told how the Turks had thrown biscuits and water-bottles out to them. The other man lay there for four days and wrote with charcoal on a board that he was hit by the cap of a shell and unable to move. But they could not get him. One morning he had disappeared. The Turks must have taken him into their trench ... Desultory firing all along the line. Shrapnel has just killed one of the 6th Light Horse machine-gunners and wounded one of our own gunners. Several of the cruisers are roaring away down on the bay. Their heavy naval shells are surging overhead with the roar of an express train. Muffled thunder of many guns is continuous away down on the left flank. The Tommies down there are nearly all new hands, poor chaps. The bomb fighting is killing them.

Afternoon—Poor Bates is shot. Poor old Bates. He came over with me only a few days ago in the Huntsend. He had been wounded during the big stunt on the 28th of June. He was water-carrying this afternoon. The bullet smashed through the water-tins and pierced his stomach, I hope he will not die. It is uncannily strange, but all the men of our regiment who have been wounded and returned, have been shot again within a few days of their return. All the regiment is remarking on it. Bates is a real merry, decent little Australian.

Evening—Poor Bates has died; only this morning I was joking with him. A bugler of ours, Roberts, was shot through the stomach. He died at sundown, singing a couple of verses of "Annie Laurie." ... A bullet has just struck the embankment above my head. It split up and a fragment whizzed to my feet.

September 3rd—In the firing-line again last night: will be working in the trenches today. All's quiet, just odd bullets and a shell now and again. There is much sickness ... Rumour is flying around the trenches that the *Southland* has been torpedoed with the loss of Colonel Linton and some men. They were of the 6th Brigade. Rumour says the *Southland* reached port under her own steam ... Well, I'll be blessed! Old Gus Gaunt has just been wounded in the arm. He rushed past me holding up his arm with the blood streaming down it, his hard old face wreathed in smiles. No wonder, after months of living with the dirt and lice. One of our big seaplanes is buzzing overhead and the Turks' machine-guns at Gaba Tepe are singing a lightning *tut, tut, tut, tut-tut-tut-tut-tut-tut*.

…In the support trenches last night. Very heavy firing came from away out towards the English lines at Achi Baba. All the warships in the world seemed to be lined up in the distance. Their song was a continual throbbing roar. It was just like a sea of waves dashing against a rock cliff to surge sullenly back only to hurl forward again with a titanic roar. I, with a few others, am detailed for duty up at Lone Pine. We'll get plenty of bombs there … We were lined up just now, we chaps for Lone Pine. The boys off duty in the dugouts called jokes to me. It's my turn to be "outed." All the other wounded ones that returned with me, have been "knocked."

Removing wounded from below Pope's Hill after the charge of the 1st Light Horse Brigade, August 7th 1915.

8

September — At Lone Pine — after a tedious walk along narrow saps then through a tunnel timbered with beams. We stumbled in the darkness instinctively ducking our heads only to thud into the wall of the tunnel where it twisted and turned. The floor was uneven with puddle holes of putrid water. Of course, no one dare strike a light; we were going to the most dangerous spot of the whole Gallipoli line. The route smelt like a cavern dug in a graveyard, where the people are not even in their coffins. We are right in Lone Pine now and the stench is just awful; the dead men, Turks and Australians, are lying buried and half-buried in the trench bottom, in the sides of the trench, and built up into the parapet. They have made the sandbags all greasy. The flies hum in a bee-like cloud. I understand now why men can only live in this portion of the trenches for forty-eight hours at a stretch...

The first Turkish sap is only fifteen feet away; by peering from behind our parapet we can just see into the inner edge of its broken bags, pierced with bomb splintered shafts of timber, and rags of dead men's uniforms. The Turks cannot hold that sap, nor can we, for both sides can rain bombs into it and make it certain death in a matter of seconds. But the Turks (game men) sneak up it during the night, throw a shower of bombs across into our trench, then scurry wildly back ere vengeance overtake them. Behind that No Man's Sap are lines and lines of trenches stretching one behind the other, most of them heavily timbered and roofed. Dead men, sun-dried, lie all between the trenches. A dreary outlook, it seems the end of the world. Bullets hum ceaselessly.

Our trench is treacherously narrow, twisty and deep — rugged witness to the haste and depth our men had to dig in seeking shelter from the bombs. The trench was once roofed with beams and sandbags, but all available timber has long since been blown to fragments by the bombs. The front wall of our trench is dug into at a distance of every few feet into firing-possies, in which two men can just stand. The trench proper, which runs behind the possies, is two feet deeper. So that if a bomb falls in your possy you kick it back down into the trench and throw a sandbag over it, then crouch back in the possy all in the one automatic motion as it were, praying that the deepening of the trench behind will shield you from the flying fragments. Narrow walls of earth are left standing between each firing-possy. These walls partially protect the men in the next possy when a bomb falls in yours. Men are killed here every hour. But if some precautions were not taken no one could live here at all. I'm handling a periscope rifle now, it's the first time I've used one. The opposing trenches are so close that loopholes are useless to either side. Any loophole opened in daylight means an instant stream of bullets. So Jacko uses his periscope rifle and we reply with ours. A periscope is an invention of ingenious simplicity, painstakingly thought out by man so that he can shoot the otherwise invisible fellow while remaining safely invisible himself. Attached to the rifle-butt is a short framework in which two small looking-glasses are inserted, one glass at such a height that it is looking above the sandbags while your head, as you peer into the lower glass is a foot below the sandbags. The top glass reflects to the lower glass a view of the enemy trenches out over the top of the parapet. It is a cunning idea, simple and deadly, but I feel clumsy with it at first. My mate is on the watch now, while I'm scribbling. So long as one man out of every two is gazing out over the trenches (through the periscope generally) the other can stand by — while it's daylight, that is.

…I'm blest if a Turk didn't unconcernedly walk down a ravine five hundred yards away, driving a little pack mule — another Johnno sauntering behind him. How unconventional! — they must have thought that with these deadly trenches occupying our attention no one would ever notice men five hundred yards behind the lines. I quickly trained the periscope rifle towards them; the dashed thing felt very wobbly and I had to crouch right back to the extreme edge of the firing-possy. Then, reflected in the wee mirror, I watched two Turks leisurely walk out of a sap. I was so staggered at their cheek that I gazed a while, then crack! — the bullet spurted the dust directly between them. They sprang up and back into the sap as if they had been shot. But the blasted periscope frame had kicked me on the jaw and nearly knocked me back down the trench. Then my mate spotted another Turk casually strolling along. Crack! — we did not see where the bullet hit, nor where Johnny vanished to. Since then, we can see Turks' heads bobbing as they crouch and run through that sap.

If the Turks facing the old 5th where we have steel loopholes and our own good rifles were only a quarter as game, or rather as foolhardy, we would put up some record sniping tallies. My mate and I got quite perked up for a while, firing away out above the dead men to those lively Johnnos behind their front-line trenches. I suppose they owe their false security to the fact that these trenches are constantly being filled by a stream of new men whose attention is immediately occupied by the trenches in front, allowing them no time to train their glasses on to the drab distance behind. Or perhaps not all our chaps have the sniper's curiosity…

These flies are awful! It is comical seeing the new men trying to stick it out. Each old hand is given a new hand as a mate, to "break in." They are going to have a rough breaking in. I can hear one chap vomiting from the smell away down the trench already. They stick desperately to their firing-possies, trying to peer out through the periscopes and so keep their attention away from the crawly things about them. My little fair-skinned mate shivers every time a maggot falls on him.

…Twenty raiding Turks rushed these first three firing possies last night. Forlorn hope! Fancy trying to surprise men whose hearts are in their mouths, whose every nerve is strained as they stand with tautened muscles, their bayonets thrust waiting at the trench parapet. Nineteen of the Turks were instantly killed in a mad rage of overstrained nerves. The last Turk fell head over heels into the trench and the sergeant snatched him by the throat, but some fool instantly blew the Turk's head off. He might have given information.

…We have just been chuckling over a bit of fun away up at Quinn's Post. The boys rigged up quite an inviting bull's-eye and waved it above the trench. Each time the Turks got a bull, the boys would mark a bull. For an outer the boys marked an outer, for a miss they yelled derision. The Turks laughed loudly and blazed away like sports. After a while an officer came along and of course had to be a spoil-sport.

…They are gossiping now of one of the every-day game things that are never noticed. There was one of the little local charges farther up the line. Our men were badly cut up. A man in safety down in the support trench saw his two mates fall close to the parapet. He jumped up and ran out under a furious fire. He brought one man in and ran back for the second, but while bending over him was shot in the back. Though in dreadful pain he dragged the other man to the parapet and both fell headlong back into the trench … One of the 12th Light Horse has just been blown to pieces by a bomb. Poor chap. He was "broken in" all right.

…Maggots are falling into the trench now. They are not the squashy yellow ones; they are big brown hairy ones. They tumble out of the sun-dried cracks in the possy walls. The sun warms them I suppose. It is beastly … We have just had "dinner." My new mate was sick and couldn't eat. I tried to, and would have but for the flies. I had biscuits and a tin of jam. But immediately I opened the tin the flies rushed the jam. They buzzed like swarming bees. They swarmed that jam, all fighting amongst themselves. I wrapped my overcoat over the tin and gouged out the flies, then spread the biscuit, held my hand over it, and drew the biscuit out of the coat. But a lot of the flies flew into my mouth and beat about inside. Finally I threw the tin over the parapet. I nearly howled with rage … I feel so sulky I could chew everything to pieces. Of all the bastards of places this is the greatest bastard in the world. And a dead man's boot in the firing-possy has been dripping grease on my overcoat and the coat will stink forever.

…This is the most infernally uncomfortable line of trenches we have ever been in, which is saying some for the regiment. We are in "reliefs" now, "resting" about fifty yards back of the firing-trench. For a couple of hours, to rest our nerves, they say. There are forty-eight of us in this particular spot, just an eighteen-inch-wide trench with iron overhead supports sandbagged as protection against bombs. We are supposed to be "sleeping," preparatory to our next watch. Sleeping! Hell and Tommy! Maggots are crawling down the trench; it stinks like an unburied graveyard; it is dark; the air is stagnant; some of the new hands are violently sick from watching us trying to eat. We are so crowded that I can hardly write in the diary even. My mates look like shadow men crouching expectantly in hell. Bombs are crashing outside, and—the night has come! If they hadn't been silly enough to tell us to sleep if we could I don't suppose we would have minded. The roof of this dashed possy is intermixed with dead men who were chucked up on the parapet to give the living a chance from the bullets while the trench was being dug. What ho, for the Glories of War!

The main track leading up to the deadly Shrapnel Gully at Anzac.

9

September—Evening—First Watch. We are back the firing-possies. The gloaming has brought the mortar-bombs flying about. They explode with a rending crash. The bullets are much more plentiful. It will soon be too dark, and we will be too busy for me to write in diary. No bombs have yet fallen in this particular portion of the trench; I dashed near pray they won't. The Turks are only a few yards away.

...*A few days later*—I'm in a hospital-ship again, let me see if I can remember things and write them just they happened. With the last rays of the sun, I was staring through the periscope for any sign of the living among the bodies. There are little khaki heaps of bodies, then twos and threes here and there lying among the Turks. Some are only rotting khaki without either shape or form. The boots last the longest. Within a few yards of my periscope lay a tale telling how furiously both sides died. The Australian's bayonet is sticking, rusty and black, six inches through the Turk's back. One hand is gripping the Turk's throat, while even now you can see the Turk's teeth fastened through what was the boy's wrist. The Turk's bayonet is jammed through the boy's stomach and one hand is clenched, claw-like, across the Australian's face. I wonder will they fight if there is an after world.

Well, the dark came, bringing a vicious increase of rifle-fire. The top layer of our possy was only one sandbag thick. The bullets ripped into this, and the sand began to flow out. As the top layer of bags subsided we had to crouch lower, otherwise our heads would have been blown off. Then came one continuous screech of bullets, a piercing chorus, ceaseless throughout the night. Then the roar of bombs in earnest, exploding in front of our trench, around us, behind us, with a blinding flash and roar! and clouds of earth and smoke, and the stench of burning cloth. Soon my mate and I had to smash two apertures through our parapet so that we could peer through and shoot the shadowy bombing men. What hell was let loose outside and all around us! A bomb blew half our possy parapet in and as I was flung back my smoke-filled eyes caught a glimpse of stars far in the sky—I wished I was up there. My new mate was frightened, so he crouched down with both our overcoats folded ready to throw over any bomb that should be thrown into the trench behind us. I tried to throw the burst sandbags together as part shelter against that screaming rain, but there came a series of shattering roars that blew the whole trench parapet to hell. It started from up the right and came crashing along, bang—crash, bang—crash, bang—crash right down the line, all mixed with leaping balls of flame, it was .75 shells—they razed our Parapet to the ground and blew into the air burst sandbags and baulks of timber and bits of dead men that came flying down plop, whack, plop into the trench.

In that inferno of smoke and fumes and grizzling explosives, I whiffed distinctly the mixed odour of smashed dead men, we simply crouched, partly dazed, and I kept firing and firing and firing. There was nothing else to do. Men in the possies to right and left were falling back into the trench, some screamed, others just thumped back. New men kept coming up from supports—stumbling over the bodies—groping along the trench—whispering up to us whose places they would take.

The stretcher-bearers down there had a fearful job getting the wounded away to the pitch-black tunnel places that led away back from the line. Stretchers could not be used in those narrow twistings. The Turks, expert fighters, use a sort of sheet to carry their wounded away in. Our machine-guns right amongst us were blazing their own hell to the inferno. Outside, the night was spitting flame from the Turkish rifles, their front-line so close that burning wads hissed down by our faces. Their machine-gun possies screeched in trails of flame. The Turk was fighting hard—both sides were stretched to the limit. Our bomb-throwers stood unseen, a glowing cigarette in each man's hand—each lit fuse after fuse, throwing bomb after bomb with a sort of sighing grunt. It makes a man's shoulder-muscles ache.

At long last the relief came stumbling in; we could not see them, we could hear them down there in the dark. At last they groped along the trench and clawed up over us, no man stepping down from the firing-possy until a new man had taken his place. We survivors of the old relief crouched down there in the trench. It was an awful feeling, waiting there, staring upward expecting hurtling bombs or mad Turks jabbing down at us any moment. At last the man behind me whispered: "File off!" I stuttered the word on, and we pressed man against man, shivering in the hope that soon we would be under some sort of cover. But in that awful slowness of moving we saw hissing sparks flying over the parapet—a choking cry, "Bomb! Bomb! Christ!" I tried to jump back but the men pressed behind were new hands and did not know what to do. Poor old King was in front of me. He jumped forward, but the men ahead crouched in the blocked trench. King was on a slight incline and as the hissing thing thudded it rolled horribly towards him. I thought the end of all things had come—I threw my overcoat over it, clenched my arms across face and stomach and pressed desperately back against the men behind. Then all was a suffocation of deathly fumes—I was on my back, quite distinctly hearing the clash of bayonets as rifles thumped across me. Then followed a strange, dull silence, ears ringing like mad. King called out: "I'm wounded, boys." I called out, "So am I, Kingey," and struggled up.

Poor King had an arm and leg broken, and other wounds. Two sergeants were struggling to get him away, but in that narrow network they did not know the way. I stumbled forward, but fell over a dying man. Another man had his ankle smashed, another was groaning with a smashed back, yet another man's leg was broken. My arm was numb, I could feel warm blood trickling down my ribs. I knew it was my own blood: I felt it belonged to me. The slow rising fumes swathed the shadows of groping men, like blind things in hell. I pressed back against the trench-wall praying they would be quick getting away the wounded, and glanced fearfully upwards, expecting another bomb. What annihilation a second bull's-eye would have been! They got poor old King and the others down a black side trench at last. Then we groped through pitch darkness into a cave-like dressing shelter, the wounded hardly moaning, just holding back their agony through clenched teeth. I was not hurt much at all. The dressing-shelter was rudely cut out in the earth. I glanced instinctively at the low roof. Thank God! It was heavily timbered and tight-packed with sandbags. I crouched down on the floor to wait my turn. The doctor was working with a tiny dull light. His A.M.C. men were all shadows; every man had his back bent. We seemed to be down in the pit: coming down the tunnel was a heavy, continuous rumbling—a sound like madmen's voices muffled by thunder.

At last they temporarily fixed the wounded, got them on and moved off. We staggered out of the tunnel, tramped through a long sap, and finally emerged on the dark hillside.

What intense relief! The air fresh and cool—stars above—the bay so peaceful. The fairy lights of the hospital-ships bespoke havens of rest. As we climbed down the track I laughed at the bullets zipping about; but it was not a usual sort of laugh; I seemed to be half on earth and half somewhere else.

They took us to the main beach dressing-station where our little lot were attended to and then laid out with the rows of wounded to wait for the day. I sat by King the remainder of the night. I nearly cried sometimes—I was not hurt at all—but those hundreds of poor maimed chaps lying there on the sand were trying to help one another with a joke, a whispered word—a smile—a look.

...In the hospital-ship *Salta*. What a contrast to the *Franconia!* Long lines of clean bunks, clean tables and chairs and decks, lifts up and down the holds for bringing in the badly wounded. Actually nurses, that smile at a man, and kindly doctors. Fancy getting into a real bed at night! This ship is just heaven.

A few days later—We are having a lovely trip, we lucky ones who can move about. Gus Gaunt is aboard—he knows every nook of the ship, and has taken me to where the best eats can be got. We promenade the place regularly. Poor King is having a rough time. The nurses and doctors are kindness itself to the wounded.

September 9th—Steaming into Alexandria. Poor King died last night at twelve and was buried at sea this morning. King was always game. He was a gentlemanly sort of chap, too, and he died game.

Back in the Egyptian Government Hospital again. How strange! And I'm glad. Same doctors, same smiling nurses, same good old tucker, and unceasing attention. Gus Gaunt fills the bed next to me, all smiles. Neither of us is hurt badly, though; we won't be here long. We have been issued with parcels from the Australian Comforts Fund, and private parcels. They were jolly well appreciated by everyone.

September 20th—All we convalescents were gorged with cakes and good things at the Recreation Cricket Ground by the whites of Alexandria who have been helping the wounded ever since the war began. Mr Harrington, I believe he is the postmaster, and his friends are thought no end of by the wounded men.

September 21st—I wish the war were over. I am getting such a longing for the bush again. I saw Darby MacNamara and Trembath yesterday at the Deaconess Hospital. Both are being shipped to England.

September 29th—We are at Ras-el-Tin convalescent home now. It is a big stone place of enormous corridors, with a huge courtyard in the centre. Rules not in the least oppressive, plenty of leave, a liberal allowance of pay, and Captain Dwyer, the Adjutant, a man thoroughly liked by all the cosmopolitan hundreds. We are mostly Aussies and En Zeds; but there are quite a number of Tommies.

Adjoining us on one side is the Sultan's palace, with its smartly uniformed Egyptian soldiers full of the pomp and sneering haughtiness of the East. At night the languorous city is lit up, it comes out and lives, the stars twinkle from a deep velvet sky, the American man-o'-war is signalling with coloured lamps, all the close-packed vessels are illuminated; here and there arises a queer native song, the plaintive melody of some reed instrument. And if a man be outside, passing the big shadowed wall of some private quarters, quite likely he will hear a soft laugh from inside amongst the rose bushes.

October 19th—This is the first place where I have really seen value for the money that Australia has lavished on her wounded. The Australian

Comforts Fund have sent us easy chairs, tobacco, lollies, and pipes, and a game of indoor tennis. State School kiddies sent us little bags full of soap, toothpowder, etc. And we have received lots of parcels with socks and kind notes. It was great! Showed us after all we are not forgotten Thought out another invention while in hospital, and yesterday worried Captain Dwyer about it. To my surprise he was quite interested. Has promised to see a submarine officer about it.

Cavalry in Nazreth, Palestine.

10

December 1st—We have been quarantined. An outbreak of scarlatina. All well, now. Captain Dwyer made time drag far less heavily by getting us all sorts of games. By the way, the invention came to nothing. The submarine officer could not be found. He has disappeared on secret duty. A number of Intelligence Officers are living in the native quarters, as Kitchener used to do. Any that the Arabs find out, they slit their throats. Those chaps live fascinating lives but they must be possessed of a sort of supernatural courage.

December 4th—Medical Board declared me on Class B. Am on guard duty here now, until judged fit to return to the Peninsula I tried hard to enlist with the Composite Regiment. They are all recovered wounded men from all regiments and armies, going out to quieten the Senoussi Arabs who have started a Holy War in the Tripoli desert. Nearly all the able-bodied soldiers are now on Gallipoli.

December 5th—The city is seething with unrest. There are rumours everywhere that at Christmas the Greeks are to unite with the Arabs and murder all foreigners and Christians. They think we are getting beaten on the Peninsula. What a surprising shock awaits the mob if they tackle this walled convalescent home. We are ready.

December 10th—Alongside our frowning entrance gate there is a forbidding wall. We know that inside is a garden. A stone house towers up, the narrow windows ominously barred. We are told the place is a harem, so of course it is of interest to us. While on guard duty this morning I noticed a head of fuzzy black curls peering between the bars of a top window. I stamped my iron-shod rifle-butt on the stone flags. The head turned into a strangely pretty face, the big black eyes trying hard to gaze between the bars. I smiled hard and it brought a scarlet-cloaked Egyptian girl to peer over the shoulder of the little dark curls. I smiled—they smiled back. I sloped arms smartly and gave them a military salute. They smiled a whole lot and acknowledged the salute by the Coptic sign. I threw them a kiss. They seemed puzzled but smiled as if they realized what it meant. I threw more kisses; they nodded vigorously and made the Coptic sign and pressed hard against the bars. I beckoned them to come down into the garden and talk to me over the wall, but they shook their heads, and smiled and made the Coptic sign. Suddenly they drew back, the shutters closed, I "sloped arms" and limped sternly up and down my beat. I'm leaning against the wall now, writing in the old diary. I always carry it in the haversack.

Two hours later—A shuttered window of the big house has just been partly opened. I was watching all the time. She of the curly head was just visible. The Red Riding Hood girl was behind her and another Egyptian girl trying to squeeze a look in, or rather, out. The third girl smiled very much when I threw kisses. The window was so directly above me that I could see the henna stain on her tiny fingers, when she reached out over the curls and twinkled them through the bars. But they all made the sign across their eyes when I tried to coax them down into the garden. After a while the shutters were closed. I will be quite sorry when my forty-eight hours' guard ends The city is ominous with rumours. It is almost certain there will be trouble with the Arabs.

Next day—During the night there were two little sparks of light from the dark window of the big house. I was standing under a lamp and waved my hand. The cigarettes waved from the window. Later, in the small hours, there was a persistent little tapping coming down from that window. But if they thought I was going to scale a forty-foot wall to a shuttered window, they found I was no blooming Romeo at all … Heard sad news of the poor old 5th today.

1916

January 3rd—Detailed for Cairo to-morrow to rejoin regiment. Good-bye Alexandria. It has been quite a happy stay here: the convalescents of all units are a huge happy family, the great dark city outside, intensely interesting. We could not believe the news of the Evacuation.

January 5th—Cairo again! Am at Gezira Overseas Base. From here they draft all convalescents to their separate units. It is exhilarating to be moving about among large bodies of men again, to experience the warm comradeship of everyone, with the feeling in the air that there might be something doing. Last night a regiment of Yeomanry fully accoutred passed us on the big English bridge over the Nile, going to an unknown destination. Their heavy horses rumbled over the bridge, their scabbards rattled, bits champed, sparks flew from hooves as the silent English regiment rode by. The 1st Light Horse Brigade are soon in harness. They have already marched out en route to the Upper Nile against the Senoussi, marauding bands of whom are massing to blow up the sweet water canals.

January 6th—With the old regiment at Ma'adi, again. It was real lonely, wandering down the old familiar lines, looking for familiar faces, and saddening to find only an odd one here and there. I think the boys of my regiment were the nicest lads in the world. The regiment is filled up with reinforcements. We landed on the Peninsula nearly five hundred strong. Our casualties were eleven hundred and forty-five. As fast as the reinforcements dribbled across, they were knocked. Only two original officers survived right through the Peninsula without being casualtied away. In my own troop, there are only four old hands left, and two of them are first reinforcements who came over with us.

January 7th—Ma'adi. Same old routine, drill, etc. I wish this damned war was over.

February 13th—The Old Bird is back, happy as a lark. All the old hands are glad to see him. The whole brigade is lucky with its officers. Some are pigs; but I suppose they can't help that. The big majority however are well liked. All the brigade likes the Old Brig. By the way, he can sling a boomerang better than any white man I know. Then the 5th like Colonel Wilson. He fined me five shillings yesterday, though, for clearing out on Sunday without leave. Major Cameron is a very decent sort, too; I think all the officers of our own regiment are liked. I detest two of them; but I suppose other men think they are all right. I haven't felt much like writing up the old diary. There's nothing doing anyway, except everlasting drill and manoeuvres to prepare us for desert fighting. At night time the city Arabs sneak into camp and try to pinch our rifles. They are getting very cheeky. They dig up our spent bullets from the rifle-range and melt them down again, into bullets for us, I suppose. We've had to organize a Flying Picket against them to patrol the camp at night.

February 19th—We are getting it hot about not saluting officers in the street. A man would need an automatic arm. We have been told, too, that the disgraceful Australian soldiers will not be allowed to go to France if they do not salute officers in the streets. Bow-wow!

February 23rd—At last there is movement—bustling camps—thousands of men packing. The excitement is mostly among the reinforcements, who have never heard bullets. The old hands just smoke away and take things as a matter of course. We are going mounted this time.

February 24th—Serapeum, on the Suez Canal. We entrained at Abu-El-Ela station, for this place. Warships in the distance appear like squatting ducks in the canal. Rumour has it that the Turks are out in the desert only a day's march away.

March 5th—We heard distant guns towards Ismailia Lakes to-day, but hardly think the Turks are there. General Birdwood with the Old Brig, was around.

March 6th—Heard guns again to-day, but I don't think anything is on. There is intriguing movement of troops though.

At night—The searchlights flashing over the canal sweep eerily across the desert and then melt up into the sky.

Any old date—Sand, sand, sand, flying sand, blooming sand everywhere. Sometimes we have to sit in camp with our greatcoats over our heads. Some days it is impossible to see the length of the horse lines for flying sand.

…The Turks have not come yet, worse luck. Anything to relieve this cursed monotony and sand. Yesterday was awful. In the evening I was riding from the canal, with three led horses prancing with the pain of driving sand in their eyes. I couldn't see five yards ahead. Presently I realized we must have passed the camp, So as the horses persistently tugged to the right I let them try their luck. We pranced right into the sick-horse lines. I had been making right out into the open desert.

March 20th—Clear days, thank heavens, even if blazing hot. The Prince of Wales inspected us today. We were curious to have a look at him.

…We stared through the brazen sunlight, to-day, surprised to hear a band, to see marching across the desert the 1st Infantry Brigade. We watched them longingly as they swung smartly over the pontoon bridge across the canal, *en route* to their train. How we wish we were going to France too. Numbers of our men have volunteered for the infantry, odd ones have cleared out to stowaway with them. Anything to escape this flying sand.

March 22nd—It was queer on sentry duty this morning, watching the big ships gliding down the Canal, and being able to talk to the men with their heads out of the portholes. A mighty work, this river cut by man! Fancy, a ship of the deep sea steaming across a desert! Last night, a P.&O. boat glided past; we could feel her engines throbbing; the desert was so quiet they seemed like the heart-beats of a mammoth out of breath. She just glided by so close, so brilliantly lit up, that we almost imagined ourselves lounging on her deck-chairs. Some of the passengers coo-eed to us and shouted: "Go it, Australia!"

…Close by, is the old battlefield of Tel-el-Kebir. Remnants of buttons, bullets, bayonets and cartridge-cases are littered there, while yellowed skulls show up where the Khamseens have blown the sand away. The scurrying winds have uncovered odd bodies in an uncanny state of preservation, surely due to some chemical preservative in the sands. Several boys looked mustily young and sleeping. It was a shock to see them, so still and quiet and old. They gave me an uneasy impression that from some aloof world they were accusing me—and really I never knew they once existed. Our boys buried them deep.

March 25th—Bright weather at last, and some gift tobacco came with it, making life much more cheerful.

March 29th—A bottle of pickles, a tin of peaches and a tin of bonza golden syrup to a section (four men) arrived today from the Citizen's War Chest Fund, Sydney. What a feast-day we will have!

…A squadron of the 8th Regiment had an interesting patrol out to attack a party of Austrian engineers engaged on the water-cisterns at Wady um Muksheib. The column rode over eighty miles of desert. The horses finished fresh, but the camels knocked up. The men are still discussing a hairy Bedouin camel-tender they rode upon. He wolfed biscuits like a dog. He had been entirely alone with his camels for two months, living on camels' milk only.

March 31st—B Squadron returned. The weather is fine. Less night duties now. All are as happy as it is possible to be—the boys are such a comradely lot.

…Two days ago, a long line of infantry staggered across the desert towards Serapeum on a route march. Men were perishing of thirst, all were in a terrible state; no water; the desert blazing; all men with heavy packs up. Reliefs were rushed out to them. Men were lying far across the desert. We heard that one officer blew his brains out. Today, some of the exhausted men showed us their tongues, covered with blisters.

April 1st—Last night a troop of us were stationed on top of a sand-ridge. The night was cold: the stars looked like chilled steel. The desert at night is utter silence: a man's mind sees shadows moving. We of the first watch rolled up in our greatcoats and slept. The sentry prodded us awake with his rifle. We were completely buried by the fine sand the wind had blown over us. As our officer remarked: "A man will never need a grave dug if he is shot in this desert."

11

April 2nd — They say that some of the infantry on that route march have died. Many are in hospital.

April 4th. 9.30 a.m. — We of the baggage-guard have just left Ismailia. The regiment is riding away. Our carriage is shared by Australian soldiers and Arabs. The Arabs are pointedly cheeky and make a hostile row. We stopped that within reason, and when a crowd of Arabs defiantly marched up our end we marched them back again, lively. They don't like it. But the angrier they get the better we like it. We have given them a square deal and half the carriage, but insist they must keep their particular brand of lice to themselves.

…We have arrived at Salhia. It looks a little Eden in the desert. The ground all green under intense cultivation, a splendid grove of date-palms, the big clean village, the minaret tower gleaming under the sun and a picturesque population of Arabs, Egyptians, and Greeks. It was from Salhia that Napoleon started on his invasion into Syria. Looks as if we are to follow in his footsteps. Guarding a stack of horse-fodder is a fresh faced young Tommy. He told us excitedly of the Arabs firing on his mate in the early hours this morning.

April 12th — Days pass uneventfully. The regiment is in strict training; no doubt we will be a formidable crowd to meet when we get into action again. The horses, too, are splendid; they stand these cursed sand-storms marvellously.

Some wild and woolly Bedouins come in from this grim desert. It was a study in nationality yesterday. My section and some En Zeds were on fatigue duty at the station when some unusually hairy camels came lurching in from the beyond. The Bedouins walked with a long, loping stride that reminded me of an emu. They were dressed in an extraordinary rough robe of goat's hair. Each wore a sheepskin water-bag slung over his shoulder. From under black cowls their jet-black eyes stared at the En Zeds as they filed silently past. You could almost hear the whites and browns say mentally: "And are you the sort of cuss we have got to fight!" The Bedouins were big men and wiry, but without boasting I feel certain our regiment could wipe out any three thousand of them, and meet them in their own country, too.

April 19th — A mixed squadron of the 8th and 9th Light Horse have made a desert raid upon the Jifjafa cisterns, where Turks with a German military artesian-plant under Austrian engineers have been digging water-cisterns. The raiders rode day and night, surprised the enemy and shot them up, capturing the camp. Jifjafa is a Turkish post some fifty-two miles out in the Sinai range.

Rumours are flying about here that we are off again any old day. The weather has been grand lately, but very hot.

April 23rd—Kantara. Early morning. We moved off from Salhia yesterday morning and today are again out in the desert. We hear the Turks are there. The Tommies at Kantara all crowded around us last night. They are such pink-cheeked, decent little chaps.

7.15 a.m.—Big news! The Turks are attacking only a few miles out. C Squadron have doubled out into the desert: we are saddling up: excitement throughout the regiment.

12 a.m.—Now I can write down what happened. The regiment hurried through the big Kantara camp, then on to a metalled road with the horses' hoofs striking hard and clear. Redoubts and barbed-wire entanglements were here and there, but soon the road ended and the open desert faced us. Then came the order: "Load Rifles!" We spread out in skirmishing order and hurried straight into the desert. The reinforcements got a wee bit excited, and anyway that same strange feeling stole over me. It always comes just when I am going into action—a curious exciting thrill, tinged with a deadly coldness. The desert spread out to the white horizon, occasional Sandhill's drab under low prickly bushes.

We got to Hill 70 where two companies of the 4th Royal Fusiliers were quickly marching into the desert. We spread out to guard their flanks. An hour passed. Then we heard the faint bang, bang of little guns. We hurried, and Lieutenant Stanfield's men in the screen in front caught several well-armed Turks. All hands gazed expectantly and then nearly at the top of a sand-rise there gleamed a few white tents and some camel-lines. We broke into a trot—the neddies were very willing. We drew rapidly closer—the camp looked strange! Then our old doctor spurred forward and we cantered by as he jumped off his horse by two dead men on the sand. Suddenly we saw that nearly all the camels were lying in grotesque positions. They were dead. And then we were certain the Tommies had attacked and captured a Turkish camp. Excitedly we gazed at the little oasis rapidly taking shape amongst the sand-dunes; soon we were cantering amongst the trees and there lay British Yeomanry horses. And this we knew: it was the Tommies who had been attacked. Right amongst us were Tommies lying among the palms, not killed, just sweating men—sunburned—very tired. Some wore bandages freshly blood-stained. We plunged past a group of dirty Arab prisoners. And under shady palms there lay wounded Tommies, gazing up at us and smiling and then a group of Turkish prisoners in bright yellow uniform and brilliant sash. Then we spurred out of the palms and my neddy leaped convulsively over the body of a huge Sudanese. I stared down at his hand clutching his crimson breast. Our horses jumped over, or sprang aside from Arabs lying on the ground, their bare brown legs all twisted up in their dirty robes. Then we cantered by the tiniest redoubt I have ever seen and around it lay the yellow uniforms of dead Turks and sprawling Arabs. We spread out and galloped over miles of sand. But the enemy had got clean away on their camels. We made contact with C Squadron who were eight miles farther out in the desert. They had the real fun, for the Turks had kept up the attack until C Squadron galloped up. All that our crowd got was a few prisoners and wounded men who had collapsed in the desert.

We are back now in Bir-el-Dueidar, the little oasis camp, tired and disappointed. The Tommies are identifying their mates. There are nineteen of them, lying in a row under the palms. They are shot through the head, and such a fine bay horse is lying by them, It is an unhappy little scene.

…A Yeomanry man is telling us dolefully that five thousand Turks attacked his brigade at El Quatia this morning. He seems to think the brigade is annihilated. El Quatia is only a few miles away. There may be something doing here then, at any time.

8.30 p.m.—Eight of our 'planes have buzzed overhead. One returned flying at a terrific pace, dropped a message, and then sped off towards Kantara.

…We can hear the faintest boom of guns.

…The wee garrison here numbered ninety-six, and are men of the Royal Scots Fusiliers. They put up a great fight. The few prisoners we have got are sullen, but one has admitted that the attacking force consisted of seven hundred camelmen.

…The fight yesterday lasted for five hours. The Turkish raid was utterly unexpected. The early hours were bitterly cold and a dense fog enveloped the desert. The Turks crept right up to the tiny redoubt. It had one sheltering strand of barbed wire, looking for all the world like a drooping grocer's-string. A cheeky terrier belonging to the sentry on duty pricked its ears and growled, then hopped up on the parapet to shield its master. The sentry could just see the terrier's stumpy tail, erect and bristly. The terrier growled furiously—the alarmed sentry siting up his rifle and challenged. A cloaked figure loomed gigantic out of the fog—the terrier snarled forward and a rifle-butt crushed him dead—the sentry fired and the shrouded figure pitched headlong down into the trench.

Such was their sudden awakening, as the Scotties told us this morning. The very fog itself that had so befriended the Turks, saved the Scotties. For the Turks and Arabs lay straight down and fired, not knowing whether they had merely run into an outpost or a heavily-manned trench. If they had just sprung forward they would have been into the tiny redoubt with its sleep-dazed men.

The fight flared up, neither side being able to see more than a yard ahead. The Arabs sneaked to the flanks for they smelt the camels. They rushed the camel-lines, bayoneting the camel-drivers. The Bedouins showed every bravery, perhaps because of the fog and the knowledge that they were creeping on a sleeping foe. However that may be, the Bedouins are this morning lying side by side with the Turks right up to that ineffectual strip of barbed wire. There are seventy-five dead desert men, and there are a few more lying away out in the desert where they fell from wounds during the getaway. The Scotties had twenty-three men killed. Their Lewis-gun was worth a battalion of men to them. The Scotties are enthusiastic about it, as they are of their commanding officer. Its moral effect when it barked in the fog must certainly have helped to hold the Turk back those precious few seconds until the men awakened.

…A troop of us rode out on patrol this morning, it was an interesting patrol; in the very early morning the hills of Sinai are soft in almost purple shadows. The sun rises above the horizon in needles of molten flame We rode on a dead Bedouin lying headlong down a gully, his face buried in the sand. He must have had just strength to crawl to the gully edge. His bandolier was empty, proving the shots he fired yesterday. His long curved dagger made a souvenir for the trooper that clambered down to the burnouse-shrouded form.

We gazed eagerly about, expectant of a shower of shots, ready to fight or gallop as the case might he. Our troop now has been heavily reinforced and we are up to full strength, thirty-three men. We are full of fight and our horses are in great nick.

But the Turk and Arab had "folded their tents and silently stolen away." We followed the broad road of camel-hooves that wound in and out among the desert hills, but the tracks ever spread away and away and away. Five aeroplanes flying low buzzed over us straight out into the desert, with the risen sun reflecting from the machines like the glint of looking-glasses.

…It is all rumours yet as to what has occurred at the oases away out on our left. The 5th Mounted Brigade (Yeomanry) the Warwicks, Worcesters, and Gloucesters, were stationed out there in the big Katia oasis area, their posts at the different oases at Romani, Katia, Oghratina and Hamisah, all a few miles apart. Five thousand Turks have attacked the scattered posts, simultaneously in the fog. The only thing certain is that of all the posts, the Scotty one at Bir-el-Dueidar is the only survivor. The others seem to be either annihilated, or their men scattered lost over the desert.

…The joy-spots of this old Bible desert are the oases. They seem to be about twenty-five to thirty miles apart, except when there are groups within a few miles of one another as in the huge Katia oasis area. Each little group of date-palms among the sandy hills shelters that most precious thing to man—water. It is in tiny wells which have been used since countless centuries before Moses.

12

April 25th—Was on outpost duty last night—shivering. While still dark, an aeroplane buzzed overhead, flying east. It must have been arctic up there … We smiled last night. A dozen Bedouin women and children camped in their goatskin tents right against our regimental camp. Two of our men were ordered to take them into our camp lest they give our position away to their menfolk. But the women were more obstinate than mules—refused to come in. At last one of our fellows picked up a child and walked towards the camp. The mother cried pitifully and ran after him. The other soldier blocked her path with levelled bayonet, but she stood with her breast against the steel and tried to kiss his hands. Our fellows swore at the women and let them go.

…The Tommies really are fools. Some of the Arabs who were shot in the attack, were fraternizing with the camel-drivers the day before and cadging food from the Tommies. The prisoners say that some of the Arabs were living in the camp here days before the attack. One of our patrols found a dying Turkish officer miles out in the desert yesterday. He says that the force which attacked this post numbered seven hundred and fifty men. No doubt the Tommies, or rather the Scotties, put up a desperate resistance. A few hours after the fight started a 'plane swooped so low over the redoubt that the sand was whirled from below her propeller. The observer shouted "Hold on—the supports are coming!" then blazed into the Turks with the machine-gun. The Bedouins sprang up from the holes they had scooped in the sand and scattered wildly while the Scotties blazed at them. When the Turks saw C Squadron galloping close up with us, and the Scots infantry coming in the rear to their right and left flanks, they rushed back to their camels and fled. They had a 15lb field-gun that the 'plane dropped a bomb on and put out of action. The Scotties swear by the 'plane.

Evidently they were in desperate need at that moment. She had to fly so low in using her machine-gun that the Scotties in the redoubt could plainly hear the *smack, smack, smack,* of the Turkish bullets through her canvas wings. The 'plane was badly damaged. The Turks drew their field-gun or guns by mules. Some of the mules shot were loaded with ammunition packed in splendidly-made iron boxes, eight shells to a box, each shell fitting into a groove. The shells were prettily made.

April 26th—We found ten more dead Turks out in the desert, dressed in their unusual yellow uniform with a red sash … There was a patrol of

Worcestershire Yeomanry with the Scotties … Was on patrol again yesterday. We found a Yeomanry man lost in the desert with a lot of camel-drivers. He was one of the men who had got away from Katia: they were mad from thirst. We have found such a lot of them now, scattered all over the sands. And numbers of riderless horses have come into camp — they gallop for the water neighing like mad things. The men and horses we find now though, are dead. Now we know what really happened to the 6th Mounted Brigade. The men at Oghratina and Katia were destroyed. They were also pushed out of Romani and Hamisah. The brigade has lost six hundred men. After we relieved Dueidar and news of the disaster trickled in, the 6th and 7th Light Horse away back at Salhia on the Canal hurried to Romani by forced marches. They found that the Yeomanry at Oghratina and Katia had died hard. Numbers of men had been bayoneted in their blankets. But many others after the first surprise had burrowed holes in the sand and fought to the death. Beside each man was a little pile of empty cartridge-cases. The 6th and 7th Light Horse were pretty mad when they found some of the wounded Tommies had been slowly choked to death. The Bedouins had twisted wire around their throats. They got the wire, a thin wire that binds the bales of horse-fodder. They tore the clothes off the wounded, first sneering "Finish British! Turks Kantara! Turks Port Said! Turks Cairo!" When the 6th and 7th got there long after the fight, they found dead men and wounded, stragglers and horses scattered for miles over the sands. Away out at Oghratina, days afterward, they got some Yeomanry wounded, still alive.

The Yeomanry officers lived pretty well: they seem to have been the sons of wealthy families. Lord Elcho was one of the captured, so the survivors tell us. We haven't seen a real lord yet.

…The horses are at last earning their feed. They stand the heavy patrol work splendidly. They are trained to the last sinew.

April 27th — Was on Listening Post last night — very cold. It is a nervy job, standing wrapped in a greatcoat, like a shrouded shadow that dare not move, staring out into the desert.

Our main outpost hill is three hundred and eighty-three, standing like a pyramid of sand three miles out from our Oasis camp.

…I am sitting on a warm Sandhill free for four hours. Writing fills in time wonderfully. This sort of active Service promises to be very interesting; so I'll explain my theories even if only to myself. Twelve of us are up here on the peak of the world; four are down the hill in a sheltered spur with the horses. Stretching before us is a sea of sand peaks. At a surprising distance away over the hills we can plainly see the tracks of Australian horse patrols, or of Bedouin camelry. About six miles to our left are the 6th and 7th Light Horse by the Romani oasis. Three miles behind us is the dark green little patch which are the palms of Dueidar, where our regiment—no not rests—is ever ready.

Now, the prize of nations, at present, is the Suez Canal, about ten miles behind us. So we, that is, our outpost, are really guarding that hundred miles of waterway with its load of ships and all that it means. That sounds comical, but is true. If the Turks come, our job is to detect them miles away. We then helio the regiment, which turns out to fight after it has helioed Hill 70, which phones back to the Canal Army and instantly the fighting machinery of an army is set in motion. Meanwhile our outpost fights. If superior numbers drive us back on the regiment, the regiment fights. If numbers are still superior, the brigade fights. If the brigade, or what might be left of it is driven back on Hill 70, then the infantry fight. While we are fighting and holding back the enemy all we can, the army behind is rushing up reinforcements, for time means everything. If Hill 70 is captured then the whole army along the Canal fights. And if the army is pushed into the canal, then England loses the Canal, and all her army in Egypt, and all her stores and her ships. She loses all Egypt and her prestige, and perhaps the very war. So now, England, all your might and power and the lives of hundreds of thousands of men might well rest on this sun-browned outpost gazing away out across the desert.

So that's that! I'll have a smoke now: I reckon I've deserved it. And the old colonel has put some sort of a stunt across the canteen funds, so that we can have a plentiful supply of tobacco.

Afternoon—This morning I explained the fate of an army, or rather two armies, and of the Canal and Egypt and England. I reckon it was good work, considering I'm only Trooper 358. I'm off watch for another four hours, so I'll explain a bit about the regiment. We are all concentrated in sections. A section is four men. A section lives together, eats together, sleeps together, fights together, and when a shell lands on it, dies together. A full troop of men has eight sections. There are four troops to a squadron, three squadrons to a regiment. I'm not going farther than the regiment. Our big world is the regiment and even then most of us don't know intimately the men out of our own squadron. Our life is just concentrated in the "section." We growl together, we swear together, we take one another's blasted horses to water, we conspire against the damned troop-sergeant together, we growl against the war and we damn the officers up hill and down dale together; we do everything together—in fact, this whole blasted war is being fought in sections. The fate of all the East at least, depends entirely upon the section.

Well now, my section is old Morry, and Stanley, and Bert Card. Gus Gaunt is a great mate but he fights and growls in another section. So that's that. Now I'm not going into details of the little troop even, let alone the squadron and the regiment. They all have their sections. Our squadron officer at present is Captain Bolingbroke, for the Old Bird has gone to France. Bolingbroke is well liked. He is a long, lanky fire-eater, fitting successor to the Old Bird. Dr Dods has also gone to France. Major Maclean is the doctor now—he is such a decent old chap the regiment has taken him to its heart.

A wind is eddying curls of dust from the Sandhill tops. Here goes for a smoke before it is my watch again.

April 28th—The Turks have all retired towards Mageibra. The 6th and 7th Light Horse patrols sent in word of burying sixty dead Yeomanry at Katia. Two days ago they found twenty other poor chaps. We are wondering as to the fate of the prisoners. The Turks must have taken some, surely. Poor devils.

…The regiment is in great spirits. There is constantly the chance of excitement in the air. We are actually getting a little bread right out here in the wilderness … The colonel must be an observant old cuss. He has introduced "spearpoint pumps." He saw them used in the Ayr sugar district, in Northern Queensland. They are very effective. They are only a pointed tube, perforated. We can hammer one into the sand and draw water from it in under a quarter of an hour. Previously, it took two men at least half a day's hard work before they reached water.

April 29th—On outpost last night towards Beetle Hill. Ideal country for snipers, but not troubled … No one would recognize the tiny Scottish redoubt now. We are turning it into a formidable position. We hope the Turks will attack, just to try it out.

April 30th—The lice are wretched damn things. A man's daily tally is about thirty except when he's right off duty and gets a chance to take his pants down, and then God knows how many he gets.

May 2nd—The regiment "Stands to!" every morning at three o'clock, a silent rising of armed men—ready. The Turks love to attack in those sleepy hours before dawn. They will get a shocking surprise if they tackle us. We are a crack regiment now—ceaseless training has made us so. A crack regiment of Australian Light Horse possess a terrible fighting-power—and instant mobility adds to our regiment the strength of two. Out in the open desert our mounted regiment could defeat two thousand Turkish infantry, and experience has taught us how they can fight. Our canny colonel will never let us be taken by surprise.

Besides the circle of redoubts we are building around the oasis, we have dismounted outposts all around, especially at night. A strong post three miles out on top of 383. By day, they can helio to the invisible 6th and 7th Regiments out at Romani and miles away back to our rear to the Hill 70 infantry garrison near the Suez Canal. By night, a ground telephone-line connects from the outposts to the regiment. And as an encircling chain, with each link silently alert, in likely dongas, wherever the enemy would most likely crawl to the attack, await our Cossack posts, grim hands ready to fight and give the alarm on the instant—then fly should the case demand.

May 5th—German taubes are a dammed nuisance. Almost daily they drone over the oasis, seeking a target to lay eggs. Our outpost on 383 generally see them miles away with their glasses. They ring up the regiment and instantly the oasis springs to life. Every man rushes his horse, leaps on and gallops straight out into the desert in a thundering scatter of six hundred horses. The taubes have never surprised us yet. They haven't scored a single casualty. And, a curious fact, they can't see us when we remain perfectly still. We have proved this from experience and captured orders. It is a curiously triumphant feeling, a feeling with a delicious little scornful thrill, holding your horse motionless and gazing up at the ominous metalled bird flying so low that you can distinctly see the hooded heads of pilot and observer gazing down, and yet though hundreds of men are watching them they can't see a single thing on which to loose their bombs—so long as we remain scattered and still. So much for the aeroplane—modern vulture of war beaten by the instinct of ground things.

May 6th—We are getting quite decent tucker here. Nothing like private life of course—but still quite a number of grateful changes from the steel-hard biscuits and bully-beef … They found the bodies of three hundred more Yeomanry lying around Katia. Hope the report is incorrect, but am afraid now it is not. We heard of it a week ago and it has not been contradicted.

13

May 7th—Bir-el-Dueidar. Was on listening-post duty last night, in a nervy position. Desert hushes all around. I did not know which way to turn. Felt a Bedouin might come sneaking up and plunge a knife in my back from behind.

…The aeroplanes that hum over the regiment at "Stand to," *en route* to scout the Turkish base look like huge black beetles droning through the darkness … One of the bombing 'planes of yesterday has not returned; another winged back to camp with the observer bleeding to death.

…An En Zed patrol met a Turkish camel-patrol this morning. An En Zed man stopped a bullet through the shoulder.

…An Egyptian engine driver away back was shot at daylight by a sniper.

…Our patrols are constantly bringing in small parties of Bedouins, women and children and old men mostly, with their flocks of scraggy goats and mangy camels. The young men are nearly all-away fighting with the Turks. These wandering Bedouins watch our desert-patrols and give information to the Turks. Strong forces of Turks are about; there may be a clash at any moment.

May 10th—Blazing sun. Countless flies.

…We have just received news of General Townsend's surrender with all his troops.

May 11th—Patrols—outposts—reconnaissances—fatigues—trench-digging—flies and sun and sand.

…We hear men of other regiments have collapsed through sunstroke. Just here, there is still a taint of dead men in the air: camels, too. A peculiar sickness is amongst us. It starts with vomiting, then a rash breaks out which in half an hour spreads all over the body. The doctor has got some stuff that eases the itch, otherwise a man would go batty. Septic sores, too, are beastly things.

…Patrols, etc., etc., regimental reconnaissances to Bayud, Mageibra, Jeffeir, Salmana, Hod-el-anna, Hamisalt. Flies—heat—sleeplessness—bad tucker—fair tucker.

…Here's a patrol "stunt," just for memory's sake:

A 'plane came buzzing in with information that a body of Turks were on Hill 462 (about twelve miles out) and that a strong party was burying stores away behind the hill. So at eight o'clock at night, A Squadron rode out, with B Squadron following four hours behind as supports. We made a detour of miles so as to attack Jacko in the rear. A screen of men was thrown out well in front, to protect the main body against surprise. Naturally enough, those men in the screen would not go to sleep—they would collect the first bullets, if any. It was moonlight, the air sweet and cool, the desert utterly still, the sand-hills in patches of moon-splashed gold accentuated by black shadows. A wilderness of stars like white diamonds glistened far up in the cold. We passed to the south of hill 383, ploughing through heavy sand. No smoking, occasional whispered talking. What worried us was the whinnying of the horses. A neddy out in the screen would whinny companionably, his mate away back in the main body would answer; other cobbers would eagerly reply until the whole column was whinnying. And the sound carried far through the desert night. The neddies are very comradely. If a horse be alone away out in the screen he will either stamp and tug in an effort to get back amongst his mates, or else throw up the sponge, hang down his head, and let the heavy sand get the better of him. But if he be in the main body or supports he will plug along side by side with his cobbers and tire himself to death rather than fall behind.

After tramping through the far-flung shadows of Hill 383 we rode into a valley of sand. Abrupt hills shadowed our right while over the low hills sloping away to the left we glimpsed a thousand black-capped tiers of peaks all sand—sand—sand. The hours dreamed along to that sighing sound of a hundred hooves plugging the sand. The hills imperceptibly closed in until they towered above us in battlements of sand. It felt queer, our sighing company riding in the darkness that rose as a wall of ink on one side, while the other was bright golden sand from the moonlight. We might have been a cavalcade riding in a valley of the moon. Morry and I were thinking what a death trap it would be if the crest above were suddenly lined with Turks.

Just after ten o'clock we rode on to a still oasis, with two wee wells. How the neddies pressed forward to drink! There has long been a likeable comradeship between each man and his horse. We spelled for two hours, then rode on out of the valley into country that resembled an ocean of titanic waves abruptly stopped in motion. During the small hours we began to droop with sleep; most of us had been on duty the night before. Men would nod in their saddles to wake with a start as their horse slid down a steep incline. Sometimes the column halted while the Heads pored over map and compass. The air grew chilly. A fog began to form; we watched it creeping around us, collecting its eerie wreaths until it closed quite silently and swallowed us.

The screen and advance guard fell back on the main body as men and horses melted into the fog. Suddenly the screen would jumble up and halt like a grotesque sculptury of mammoth shapes. The column would halt upon their haunches to find that directly under the leading horses' noses there fell a cliff of sand for hundreds of feet. There was no warning until the last two feet. After that, came a step-off not on to sand, but into hundreds of feet of fog. Then we would spread out and grope until a way was found, and down would flounder the horses, right down that floury wall up to the saddle girths in sand, their steaming breaths blowing on the rump of the horse in front. At some periods the fog was so dense that I could not distinguish the head of the horse in front of me. The screen and advance guard was, of course, no longer of any use, so they fell in with us. We rode knee to knee, we could smell the horses, we could hear their breaths and the *choick, choick, choick* of their squelching hooves. And never the sound of a rifle-shot, nor the shout of a Turkish sentry. At last we had to halt, or we might have developed into a lost patrol. This was at 4 a.m. We dismounted, and flopped in the sand beside our horses and shivered silently until at last the sun arose and so slowly pierced the fog. Unwillingly it lifted; its wisps of moisture flowered into rainbows that fainted away as the sun grew red and angry. And there, not a mile ahead of us, stood up Hill 426 like a big squat pyramid tipped with a shaven crown. Cautiously we circled it, the horses carrying on gamely for hours and hours, but never a sign of a Turk, and never a sign of Irregulars burying stores. Just another cursed "Gawk Act," caused this time by the mug 'planes. They have brought in quite a lot of dud information of late. Such does not matter to them, but it has caused us many sleepless nights. So we floundered home across the limitless, dreary sand.

May 13th—On Cossack Post last night.

May 14th—Sapping again this morning, from four to six. "Stand to" at a quarter to three now, so a man gets very, very little sleep. The sun blazes by day. The flies are so bad that at times they remind me of Lone Pine. They have been feeding on dead men we haven't found time to bury, and dead camels—we can tell by their smell. A hot wind comes from the desert and makes our "gunyahs" stifling. These shelters remind me of the gunyahs of the northern Queensland aboriginals. How I would love to be in Queensland now, if it was only in a blacks' camp. Our gunyahs are simply palm branches thrown together under the palms. We crawl inside, exactly as the blacks do into their gunyahs. The wind seems to be scorching the life out of us. And our wells are falling off in water. Each man now is only allowed one water-bottle full in twenty-four hours. The water is brackish. The medical authorities issued chlorinated tablets. Each tablet, dissolved in a bottle of water, is guaranteed to kill all germs. It does. It will also burn leather and eat the rust off stirrup-irons. Which is what we use them for … The tucker is wretched today.

May 15th—On patrol this morning long before sunrise to Hill 383. The sun rose, a ball of quivering fire, hurrying from the east a wind straight from a furnace. The horses bent their heads and gasped. A squadron of our fellows rode out last night for an all-day stunt, with one water-bottle per man. Poor chaps.

9 a.m.—Back from morning patrol. We have been slipping away to drink from the horses' well. It is forbidden. The water is awful, but a man must drink something.

1 p.m.—I think this is the most hellish wind I've experienced yet. It sears through the oasis, through our blanket shelters and scorches our naked bodies The smell of rotting camels is extraordinarily noticeable.

3.30 p.m.—B Squadron has just come in with a few Bedouins and thirty camels. Only one man collapsed from the heat.

5 p.m.—The quartermaster has just issued us with some canteen stuff. Our section share is two tins of fruit, one tin of sausages, one tin of syrup, one packet of biscuits. The lot cost thirty-four piastres. What a gorgeous blow-out we will have when it gets cool enough to eat!

May 16th—Some of B Squadron have collapsed after yesterday's patrol. They had a terrible trip.

…We were issued with two barrels of beer yesterday—half a pint a man, at two piastres (5d.). It was decent of course of the authorities to think of us, but the great majority of us wish that that particular camel had been employed in carrying us two cases of jam, in lieu of the two barrels of beer.

…On a reconnaissance to Bayud the Anzac Mounted Division, the 6th, and some of the En Zeds got hell. Rode out with only one water-bottle per man—terrific heat—forced march—some men went raving. Stragglers, half conscious in their saddles, came staggering into Katia upon open-mouthed horses whose eyes were bulging from their sockets. One of the squadrons had a hot fight with the Turks—their rifle-bolts were practically red-hot in less than twenty shots. The En Zeds felt the heat awfully too.

We have proved that we can stand heat even as well as the desert Arab; what is more, we can work in it. So the Bayud reconnaissance must have been terribly hot to drive some of our men mad. Of course it would not have happened if we had had water.

…The 12th Light Horse have arrived. I walked out in the sun to see them and had to shield my face with my hands. The burning wind seemed actually to strike me. The leather of our boots is shrinking from the blazing sand.

May 18th—The evening before last, a party of us was detailed as covering guard for the Scots Fusiliers as they marched away. They were the men who reinforced their comrades from Hill 70 during the attack on Dueidar. As they marched out of the oasis, the regiment lined up and gave them some rousing cheers. The Scotties and Tommies did not understand at first, then they were very pleased and answered back heartily. I was in charge of the rear-guard as they marched back along the desert road. I could see away on ahead, how every lighted match betrayed the marching column. There was no real danger though it was quite possible a hidden body of Turks might blaze into them. From far down ahead in the column would float the laughing voices of men who were very, very glad to be leaving the desert. They are bound for France. Then I would gaze behind for the shadowy forms of the rear-guard horsemen to move up again. When we got near Hill 70, we struck the metalled road. The heavy boots of the infantry made a song of perfect rhythm. It is very interesting to watch, or rather hear, the movements of troops in the night. Strangely, our horses were averse to the hard road. But presently they took to it with delight. Two days after, when we again faced the desert, it was hell's own job to get them off the road.

…Constant night duties make us awfully sleepy. And in our reliefs by day the flies won't let us sleep. One of our officers has been sent away, through sunstroke … We hear from the 6th Light Horse, camped now in the open desert, that fifty of their men and three officers are down with sunstroke.

14

May 20th — Bir-el-Dueidar. Last night a fool airman brought in information that two thousand Germans and seventeen hundred Turks were advancing on the post. We stood to arms from two o'clock this morning. I would have been off duty, so bang went my dreams of a few hours sleep; I cursed everything — everything!

Thank God for a cool day at last.

7.15 p.m. — Excitement in camp. Expecting an attack by those Germans. Hope it comes off — anything to break the monotony of fatigues — heat — sand and flies. Horses all saddled up; everyone ready and pleased.

…We "stood to" last night for a few hours, then slept beside our horses, rifles in hand. Of course the enemy did not come. Thank God we got some sleep. These constant false alarms will give us the jumps … Yesterday the quartermaster got in a lot of tinned stuff. Our share cost us 10s.

2 p.m. — A cool day. We have good tucker now and sufficient water.

…These aeroplane reports are "fishy." A fortnight ago, one reported that fifteen hundred Turks were advancing on railhead. It turned out to be a gang of Egyptian labourers. Generally, the enemy the 'planes report vanish into the desert air. The trouble is, we have to ride out and lose sleep over the myths.

May 21st — Our gunboats bombarded El Arish, a main Turkish base about sixty miles up the coast. The information that the bombardment was a great success, leaves us cold. It will simply wake the Turks up to the fact that, like Christmas, we are coming. They will get ready for us.

May 26th — Fatigues and duties and sleeplessness.

…Two monitors and a Sloop again bombarded El Arish, fourteen 'planes dropping bombs. The bombardment was a great success. Was it? Perhaps from the Turkish point of view, for it's a certainty they will strengthen their defences.

…The 6th and 7th Regiments have gone back to Hill 70, dead-beat.

…Our regiment has developed a marvellous precision and quickness in marching out of camp at a moment's notice in the night. Fully armed and without a sound five hundred of us will jump up and in a matter of minutes be mounted and away. Extensive training will do wonderful things. Should the enemy attack, instead of hemming us in, they will suddenly find five hundred men blazing into their exposed rear.

June 1st — The New Zealand Mounted Rifles went out to attack a

Turkish camp near Bir-el-Abd. All through the night, we rode out towards El Katia to protect their right flank. All we heard though was the guns. The En Zeds smashed a Turkish outpost but the main body cleared back into the desert and were bombed by our 'planes. We rode back last night.

11.30 a.m.—This morning, the Turk has hit back. A 'plane bombed the 3rd Regiment at Romani. The first bomb smashed the wireless, the second the wireless lieutenant, while other bombs killed eight men and wounded fifteen. Thirty-four horses were also killed. The taube then circled low and turned its machine-guns on the camp. The horses stampeded into the desert. Three hundred got away. Most of them were caught at El Katia. A few even got away back to Hill 70, while some reached here.

…What a pity the Turks won't stand up and fight. They seem to be fighting of late months on the principle of skirmishes and quick getaways. The New Zealanders blame their horses for whinnying the other day and so putting them away, after all their weary marching.

June 3rd—The 12th Light Horse have marched back towards the Canal. The Highland Light Infantry have taken their place. The H.L.I. are men in kilts. They wear no pants. Jolly cool and convenient, we imagine. They won't get any lice in the seams of their pants anyway. The weather is cool. Had a very welcome bread issue; and some "bucksheish" comforts arrived the other day—four tins of fruit, tobacco and cigarettes to every section. From the Overseas Fund, I believe. Good luck to them. We are getting much more stuff from the quartermaster's canteen now, for which we are jolly grateful. But it is expensive: each man only draws two bob a day. There is plenty of water now, thank goodness. Still, I could not keep out of trouble, somehow or other. A bombastic sergeant put the rule over me for not shaving. I was too dashed worn out for want of sleep to care what he did, so I told him my opinion of him in sultry desert language. Captain Bolingbroke came to my rescue and settled the matter or I would have got it pretty hot. It was decent of the captain to stick up for me, but for the time being the sergeant wins. He has put me on an extra picket duty.

June 4th—I went to sleep standing up last night, leaning on the rifle. Simply couldn't help it. The 3rd Light Horse casualties from the 'plane bombing was thirty-five men and numerous horses.

June 8th—Am on guard at the canteen to-day. The poor Scotties are hurrying up five at a time, the leading man with an Egyptian pound-note in his hand and the wealth of the Pharaohs in his eyes. They are only allowed 4s. 2d. per fortnight. I'll never believe a Scotsman is "tight" after this. The poor beggars simply haven't got it. We are millionaires to them. They spend their fortnight's money on one terrific bender, two tins of peaches and two tins of pineapple.

…We have dug down to a big Roman bath and an old Roman well. There are metal remains of spear- and arrow-heads. So this was an old Roman camp. I wonder how much night duty the old centurions had to do!

…There are big movements of troops out to El Katia. Perhaps at last there may be something doing … Our old colonel is a shrewd old cuss. He's up to all sorts of cute dodges to deceive the 'planes that fly overhead. No wonder they think we are not worth bombing. If ever we are attacked too, the enemy will be deceived at many points … Taubes bombed Kantara yesterday and killed some men and horses. Several of the taubes, when chased back by our own 'planes, dived low over the oases and spitefully rattled their machine-guns at us We have an idea that the taubes are superior to our aircraft. They are quicker and better machines.

June 14th—The heavens have fallen, surely! The authorities are granting us leave to Port Said, for forty-eight hours! Three men out of each troop are to go at a time. What ho, she bumps!

June 18th—Seven of us made up one Standing Patrol last night, about two miles out in the desert. During my watch I was holding the seven horses. They were very sleepy, standing so still! One old bay's eyes slowly closed, his head sagged, his knees bent, and if I had not jerked the reins he would have lain down. A pretty little brown horse stared imploringly with big black eyes that slowly closed in the moonlight. But I would not let him lie down. The other neddies' heads drooped as tired as anything, one snored horribly, two others breathed great horse sighs. One neddy flopped straight on the ground, asleep. I did not have the heart to jerk him to his feet. A baldy-faced old chestnut watched events with one eye half open, and then seeing nothing happened, flopped down beside his mate. And the rough-maned little brown horse, sighing heavily, sank to his knees. They were all asleep except me, who was nearly so, intensely envying the animals their irresponsibility. And all my mates lay huddled asleep on the sand except big Morry, a veritable giant in his greatcoat, standing, clear in the moonlight like a sphinx, gazing out into the desert. And my ears were listening, my eyes staring. The desert dogs came creeping close, and howled.

June 19th—Our 'planes flew overhead to bomb Jacko at El Arish. He will retaliate by laying eggs on us … The Old Brig, said "good-bye" for six weeks this afternoon. He is off to England on some damn parliamentary business.

June 20th—A report is flying about that the Turks got one of our 'planes, and damaged two others. Our 'planes are said to have got seven of the Turkish machines, and bombed their aerodrome.

June 21st—Yesterday, fifteen bags of mail (thousands of letters) came for the 5th. And yet there was not one solitary letter, parcel, or even newspaper for me.

June 22nd—Morry, Stan, Bert and I are going into Port Said for forty-eight hours leave. Heavens, what a time we will have. No night duty, no fatigues, no "Stand to," no cursed desert for nearly two nights and days. What luck that we knocked off buying canteen stuff and saved our few piastres!

Surf-bathing!

15

June 23rd—Port Said. Early morning—and just out of a shower-bath. Heavens! the world rains miracles.

We are in the Continental Hotel. Grand rooms. I awoke to see masts of ships and to hear the sweet music of street traffic. But I could not sleep well even for those few hours. The bed was too strange. I kept waking up and thinking I was on sortie duty. When we strolled out Stan and Bert bought khaki shirts. They looked utterly changed, clean and cool. No doubt they are "the boys" of the section. Morry and I looked like nothing on earth in our desert-stained old tunics. We felt ashamed to walk beside them. Port Said is cleaner than Cairo.

I'm damned if we weren't stopped by two English military police armed with cudgels. They demanded our passes. We showed it them, They told us then that we were not properly dressed, for two of us wore khaki shirts instead of tunics. We had an argument, of course. But it's not the fault of the police. Here we are for forty-eight precious hours, after months in the rotten desert fighting England's battles. For months we have not changed our clothes; we have lived in them, slept in them, always ready for instant action. And in these few hours of leave we buy cool khaki clothes and discard for the time being our old clothes thick with sweat and grease and crawling with lice. And the cursed English military law says: "Keep those vermin-infested clothes on and go out into the desert with them again." Well, we won't! So that's that!

…The business people here charge us exorbitant prices for everything, of course.

…After a pleasant wandering, we hired a boat and sailed down the mouth of the canal … Dinner at the hotel was good and pleasantly served. We jubilated with two bottles of wine and did our best to forget that war ever existed.

9.30 a.m.—Surf-bathing this afternoon, Surf-bathing! It's a wonder the sea did not take a fit … We found trouble again when returning to the hotel. Stan and Bert had changed into their clean clothes and our luck was stiff enough to run into two military police down a side street. An argument, of course, and Bert flatly refused to put his tunic on. But one of the police spoke very decently and explained it was not their fault about the thick old clothes, it was just military rule; which of course we understood.

But as Bert said, that is no good to us. The military Heads don't have to wear our filthy rags. We had a long argument with the police, but I'm too disgusted to put it down.

11.30 am.—Last night after dinner we went to the Kursall. Now it was nine o'clock, Stan and Bert thought surely it wouldn't matter to go to the Kursall without the old tunics. Stan even put a khaki tie on. He looked quite smart in his new clothes and curly hair. I kept gazing at him; could hardly believe he was the dirty desert rat I'd worked with all these months. We trooped light-heartedly downstairs and walked right into a Sergeant with a squad of military police. We were immediately ordered to go back into the hotel and stay there, or else put our tunics on, otherwise we would go straight to the guard-room. We could make our own choice.

June 25th—Yesterday morning we went surf-bathing, had a drive around town, and in the afternoon a look over H.M.S. Jupiter, then a sail down the canal, thus ending a delightful little holiday. then came the third-class trip in the train, then the long dreary night ride and now Dueidar, with to-night on night duty and all that the desert means again.

June 27th—Sergeant Edgington came up to Stan last night and ordered him to report to Orderly Room in the morning, to answer a charge of being improperly dressed in Port Said, as charged by the military police. It was a shock to us.

…The colonel dismissed Stan's case. I'm blessed if I know why the M.P.'s have such a set on us fellows. They are criming the Australians for the most trivial things. One of our own regiment was crimed because he had sewn a button on his tunic pocket!

…We hear that the police are putting in lying charges against our chaps on leave now. I wonder why on earth they have got us "set."

…This morning, several of us spotted an aircraft very high. Then the sun shone on her—white! The long-drawn shout rang from lip to lip "Taub-e!" "Taub-e !" "Taub-e !" Then *tut-tut-tut tuttuttuttut,* as a 12th machine-gun opened fire. The whole camp sprang to their feet gazing warily up through the palms into the clear blue sky whence came that peculiar droning noise as the white bird sailed serenely high. The camp resumed its normal life. Half an hour afterwards field-glasses were again levelled into the sky, little groups of officers and men gazing from the edges of the palms. Suddenly the colonel shouted "File out!" and blew quick blasts on his alarm whistle.

The camp just flew! Every man leapt on his horse and then from behind every palm they tore out into the desert, all bareback, many men merely using the halter his horse had been tied up with. They galloped and spread out in all directions like a web with the oasis the spider, then remained perfectly still gazing up while from all the redoubts stuttered that brazen *tut-tut-tut tututtuttuttuttuttut* while the white bird droned closer and menacingly large right overhead. Heavy rifle-fire broke out from the redoubts, as the whirring bird circled and doubled low over the palms. Then high up came a heavy buzzing and one of our own 'planes tore out of the sky directly for that damned wasp. She headed for Romani and soon we could only see our own fast disappearing 'plane and could just catch a faint buzz, it was one of the little daily events that keep us alive.

July 2nd—'Planes continually annoy us, but we are too well organized now for them to do much harm. When a taube approaches us, Romani or El Katia miles away, the phone immediately sounds away back in Kantara, on the Canal, and the whole army for a hundred miles knows the enemy 'planes are on the war-path. Our own 'planes rush out—but don't always win.

…Our old doctor is a sympathetic sort. Put me off duty for a couple of days so I can get some sleep. Gave me pills that will deaden me, so that I won't be jumping up every four hours to go on relief.

…Our old colonel got even with the Port Said military police. The regiment went dashed near hysterical with delight. The colonel found that the commandant at Port Said had no jurisdiction to deal with cases and could be made to remit the cases to the soldiers' C.O. in the Field. So when the military police grabbed all of a batch of our men on holiday, the colonel insisted that every case must be tried by him at Dueidar, and furthermore that the M.P.'s must come out into the desert to give evidence against the men they charged.

You cannot understand our delight as we got ready the horses for that crowd of police. We picked every rogue horse in the regiment and no doubt some of the devils are buckjumpers. How we laughed in anticipation as the horse-holders rode off to Gilban.

Well, the police got off the train, and there were our men waiting with the horses. The police were nervous. Many of them had never been on a horse before. Our men held the indignant horses while the police mounted. Some mounted them from the wrong side. Quite a number caught hold of their horses' ears to climb on, while one man actually caught hold of a tail and nearly fainted when the horse snorted.

Well, they were all finally mounted, and our men, their duty done, let the horses go. They went! Straight out for the desert in a mad gallop of flying hooves, tossing manes, and snorting freedom. Military police were scattered all over the desert. Only two succeeded in sticking their horses. Our horse-holders nearly made themselves sick with laughter. The police looked very sheepish picking themselves up off the sand. Some men clung on for a mile and had to walk back. The horses never stopped until they galloped neighing in exuberance right into Dueidar. Our regiment, at least, expect no more trouble from the Port Said military police.

July 4th—We've had a little stunt that was very well managed. Late the afternoon before yesterday (I haven't had a chance to write since then) a 'plane brought information to Kantara that a hundred and fifty camels and some irregulars were watering out at a desert well by Jeiffer. The regiment moved off at ten at night. We were pleased when we were told the plan, which was simply to make a wide circuit and come in behind the Turks before dawn. Numerous other stunts of ours have been betrayed at the last moment. A desert dog would bark and away would go the Turks. But by this encircling movement, a similar getaway would bring them right into us.

The pitch-dark night ride was similar to most other desert patrols. We came on the well just at dawn, and closed around it. No enemy were there, but their tracks were. In the rapidly increasing dawn the regiment split up into three columns, C Squadron hurrying off on the tracks, the others scattering out for any sign of these men who can disappear like a breath in the wind. Then in the pale light came a heavy buzzing as of a prehistoric bat which circled over us, to dart off after C Squadron. A mile away and we could just see her. She dived like a hawk. *"Boom! Boom!"* then the rattle of her machine-gun. The Bedouins and camels stampeded but the 'plane circled around and around the camels just like a stockman around a frightened crowd of sheep and she kept the mob circling around and around one another, and machine-gunned the obstinate ones that persisted in breaking away and so held the panic-stricken, grunting brutes until C Squadron galloped up. It was a clever little piece of 'plane work.

We got one hundred and nine live camels but only few Bedouins. They had scattered like chaff before the wind.

We returned well satisfied for there really had been something in this stunt. And it was well managed. That particular part of the desert was very rough. None of us had gone over it before; there were no land marks; even the well had no oasis. Jeiffer simply means a "well in the sand." And we arrived at the exact spot to the minute.

Our poor game neddies marched straight back to Dueidar. They were still full of pep even though they had done forty miles over sand without one drink.

16

July 5th—The Port Said police have sent word that the 5th may go to hell before they break in any more buckjumpers. Good. That means they daren't override our men on leave. Quite an army of Gyppo labourers are working on a desert railway-line out towards Romani.

June 6th—Stan got a bonza parcel by post. All tinned fruit stuffs, I share his luck—mine's out.

June 7th—The Provost Marshal has sent out sheafs of orders as to what the troops have not to do when on leave in Port Said. He can go to hell, or mount for another desert ride.

…The old regiment has its tail up—great tucker yesterday morning—fresh steak and bacon for breakfast; the stew at night had meat and potatoes in it! There is plenty for all. This morning the orderly sergeant actually came around inquiring "Any complaints."

July 10th—The brigade has indulged in "an affair of outposts."

July 11th—Good heavens! Rissoles for breakfast. The war must have ended!

July 12th—This cursed desert monotony is nerve racking: we are the Lost Legion. Some of the Light Horse have actually managed to smuggle right away to Port Said and there stowaway on transports *en route* to France. Many of us have tried hard to get away legitimately.

Curse this inaction.

July 14th—Bert rode across to Romani. His cobber in the 2nd Light Horse told him they had only just finished finding and burying the last sun-dried poor old Yeomanry scattered around Oghratina and Katia. Numbers had their skulls smashed in. Where the machine-guns had been firing, the Yeomanry gun-crews lay dead in huddled heaps, the machine-gun belts full of hundreds of empty cartridge cases. Fronting each gun were little heaps of Turks and Bedouins. Our men tracked a Tommy sergeant. He had crawled away but was too weak to mount his horse. But he had taken its saddle and bridle off to give it a chance, then lay down and died. For miles over the desert the sands told of hundreds of lonely fights right to the bitter end.

July 18th—The 3rd Scottish Horse relieved the 12th Light Horse at Dueidar yesterday.

…A long camel-train came along carrying some battalion's baggage. A considerable quantity of it seemed to be cane chairs and deal tables and folding bunks for officers. Which makes us remember the days when we were perishing out here but could get no camels to carry water to us. Our own officers do without furniture anyway. They are satisfied to heap up a clump of sand for a table until they can pinch a box from Kantara.

July 20th. 8 a.m.—Last night our squadron were to have gone out on a "Gawk Act." Most of us were grumbling, as usual, when the camp was electrified by the news that three thousand Turks had marched on Bir-el-Abd, three thousand at Bayud, two thousand at Jameil, battalions and thousands of camels marching along the desert route. The men received the news jeeringly, but shortly came orders for the regiment to be ready to file out at a moment's notice.

…The 6th and 7th Regiment patrols have engaged the enemy screen advancing at Oghratina … Flying patrols were shot back by strong bodies of Turks east and south of Hod Abu Rodha. Excitement! the chance of a fight at last! Marvellous how the air cleared! The jeering sarcasm in memory of our false alarms vanished. Whistling and singing broke out all down among the palms.

The time came to file out, but the order was "Stand Fast!" and later we turned in with all equipment on.

…At half-past two this morning the regiment moved out but we are back in camp again, anxiously awaiting confirmatory news of the Turkish advance. The old regiment is like a boy—cheerily whistling, but heartsick with fear of another false alarm.

10 a.m.—A 'plane has buzzed in with news—the Turkish infantry are rolling in upon Oghratina—the 2nd Light Horse Brigade attacked and were counter-attacked. Just before dawn yesterday the 6th advance screen rode right into the advancing infantry. The Turks charged wildly snatching at the horsemen's bridles, but the 6th wheeled and galloped over them shouting and hammering down with their rifle-butts. Both the 1st and 2nd Brigades have now clashed with Jacko with the exception of our regiment, and we are mad that we are out of it. It appears that Major-General Chaytor, the New Zealand General, flew from Romani over El Abd and discovered the marching advance guard of a Turkish army. What ho! She humps. There'll be something doing at last. Rumour tells us that the Turkish advance guard are twenty thousand strong.

…Our own 'planes are continuously hurrying overhead towards Romani and Oghratina. We are ready to move at a moment's notice. We've missed the first scrapping. Captain McNeill was with them on special duty—We listened eagerly to particulars. The Wellington Mounted Rifles got into it too at Oghratina, the Wellingtons have taken our place in the brigade. The Turks have dug themselves in there and are advancing on all the surrounding oases. They are eighteen thousand strong at Oghratina;

rumour says other divisions are coming a day's march behind. The old regiment feels like a butcher's pup expecting a feed.

July 21st—Was on standing patrol last night—we were all ears, eyes, and anticipation. From miles away like a sigh in the night we could hear a faint outburst of rifle-fire. A desert dog came sneaking among the dry leaves in the oasis and startled the horses—we gasped, then cursed and turned our faces to the east again, listening—listening. We welcomed the fiery dawn. On returning to camp we passed the regiment filing out, horses fit and contented, men in tingling humour, ready for anything. There are numerous reinforcements among them who have never been under fire. It is arousing to watch the staring look in their eyes when faintly they catch the *boom-oom-oooomm* rolling in from away out towards Oghratina, Hamisah, Hill 245 and Sagia. When the breeze comes in desert-puffs it carries the sound a surprising distance. I grinned when a new hand turned and said to his mate: "Why, Dick, hear that? Fancy a storm in the desert!"

Trainloads of Scottish infantry are rumbling out to Romani by the desert line. 'Planes, flying at a great rate, are continually coming and going overhead The nasty snouts of guns poked from a trainload that has just thundered along the line. Things are getting interesting. A man will be experiencing that old queer feeling next. The horses know perfectly well that "something's doing!"

5 p.m.—Clashing of big patrols along the line for miles to-day—whole brigades are patrolling now. Casualties are coming in. The Turks are advancing over miles of country, digging themselves in preparatory to advancing again. The Wellingtons have been heavily engaged. Our horsemen are checking Jacko from daylight to dawn—steadily retiring— fighting every yard of the ground, a whirlwind of flying tactics. Troops are coming—coming—coming.

July 22nd. 3 p.m.—"Stand to" at half-past one this morning. A Squadron then rode out to Hill 383, placed outposts, sent out wandering patrols, then immediately the rest of us, dead tired, stretched out on the sand fast asleep. Frantic alarm whistles woke us and we galloped for our lives from a taube.

…We've been issued with the blasted hard old biscuits again The boys are just wild to have a go at old Jacko in the open. We know that he is the equal of any soldiers in the world at trench and bomb fighting, but there is a spirit of utter confidence among all of us that we are individually better men. We will soon know.

July 24th—Yesterday, very early, we rode out to Sumara Oasis. We were queerly shocked to find there were no Turks there. We heard them though, to the east. *Boom—oom—oomm! Boom—oomm—oomm!* then the rattle of musketry and the *rut tut tut-tuttut-tutut* as the machine-guns formed the chorus. The sharp rattling of the rifle-fire in the clear desert air sounded different to the rattle in the closed-in trenches of Gallipoli. Presently the reinforcements were exclaiming excitedly at the pretty little puffs bursting in the air. Then they saw the shrapnel bursting seemingly perilously close to the buzzing 'planes. The regiment was just eager for fight, but we didn't get it. It appears we are so damned efficient that we are a mobile regiment, which means that we can act quite independently of the brigade, and General Chaytor is keeping us by to rush us into any hard pressed point when the attack develops. H'm.

…The regiment has returned, the section is on outpost in Hill 383 now and so I have a chance to write up the last few days in the old diary. The Wellington Mounted Rifles have taken our place in the 2nd Brigade. The 1st and 2nd Brigades have constant day and night fighting; they take it in turns to ride about ten miles out from Romani and fight for twenty-four hours, the Turks gradually pushing them back. They then ride back to Romani for a few hours' sleep then ride out again. Each night now sees exciting bayonet fighting in the palm oases, the advancing infantry, their footsteps muffled by the sand, often pass on either side of troops of our chaps—men in groups have fought jumping back with the bayonet, the horse-holders leading the horses away through the gloom—which the Turks come too heavily it means a wild rush for the horses; the Turks have even grabbed the bridle-reins of some of our patrols. The Turks have night snipers out as well as the regulars by day.

Chauvel's plan seems to be to lure the Turks right between Dueidar and Romani, and perhaps on to the infantry redoubts. We are beginning to appreciate the plan.

…The guns are booming much nearer Romani and every now and again comes an increasing clutter of rifle-fire—Jacko is drawing in close!

July 25th. 8 a.m.—We few left in the oasis, have been chased out of it three times in thirty minutes by these damned taubes. I wish their engines would bust. A man's nerves are at a pretty lively strain, and these damn things come and play music on them.

Boom! Boom! Boom! Boom! Bombs at Romani.

…One of our 'planes has been brought down.

…There was heavy night and dawn fighting again last night all along the line. The Turks dig in by day—our fellows worry them—Jacko advances again by night—and so on … Jacko's 5.9's are exploding with a resounding *bang! bang! bang! bang! bang!* Prisoners tell us that the Turkish Army is composed mostly of picked Anatolian Divisions, men who fought on Gallipoli. One item should be of interest to the Yeomanry, for one of the Turkish regiments was with the raiders who wiped out the Yeomanry Brigade at Oghratina.

Jacko is driving herds of cattle with him. He may have a queer organization, but by Jove he gets there. He's done a hundred desert miles already and his big soldiers are all in first-class condition.

...Last night I again asked the old colonel to send the section out scouting in the enemy's country. He listened patiently, but refused. I hope I looked as sulky as I felt. I went back to the gunyah, answered Morry and Bert and Stan as surlily as I could, then lay down and went to sleep.

An hour later Lieutenant Stanfield was bending over me, poking me in the ribs.

"Idriess!" he said.

"'Ullo!" I growled.

"I'm off on a dangerous stunt—are you on?"

"Of course!" I said, and jumped up, wide awake.

Light Horse ready for a charge by the Turks at Nalin, in Palestine.

17

July 1916—Lieutenant Stanfield then explained that he and Lieutenant Broughton with a party of men, were going twenty miles out in front of our own Posts to form a little base where we could hide with tucker and horse feed. From there, four men and an officer were to spy on the big Turkish camp at Mageibra. Our Heads fear they may make a sudden flanking movement between Romani and Dueidar, and thus attack the hard-pressed Romani forces in the rear. Our job was to rush in news of any such enemy movement.

The party that did the actual spying, emphasized Lieutenant Stanfield, would probably be cut off and never return. I pleaded hard that Morry and Bert and Stan be allowed to go too. The lieutenant grinned as he walked away saying the men were already chosen and to pick four out of one troop would cause jealousy. So the section are ruffled. We consider we are the scouting section of the regiment and should have had the job on our own.

July 31st—How I wish I could describe the most exciting trip of my life as it all happened. When we filed out of Dueidar the boys collected in laughing groups, inquiring our address so they could send parcels when we arrived at Constantinople. They asked us what we'd particularly like. One fathead surmised we were going to link up with Townsend. Others called to us hair-raising information about Turkish prison camps.

We rode out into the dark, a silent company, presently passed the oasis Bir-el-Abd and left the road for the open desert, steering by the stars. The Turks were active, for out where the troops were fighting the sky was pierced by Verey lights.

We expected every moment to ride on top of hidden Turks: the horses seemed to sense the tension, they plodded along so quietly. Around us broke out spasmodically sharp bursts of rifle-fire: once came floating the shouting voices of men in some mad bayonet fight. In the small hours we cautiously entered the lonely oasis of Nagid. Here we were to store our rations and horse fodder. A similar party of New Zealanders were hidden somewhere in this oasis. Very quietly we nosed about until Lieutenant Stanfield's horse tangled up in a field telephone-wire and down he thumped. We listened breathlessly, but Stanfield was up on the instant holding the struggling horse's head. We located a tiny well, watered the horses, then caught hold of the telephone-wire and ran it along and down into a secretive donga where the En Zeds were waiting. With ominously smiling faces they whispered us a welcome; they rode away *en route* back to their brigade, wishing us luck. One fresh-faced chap grinned as he mounted his horse: "It's just possible," he said distinctly, "that we'll meet again in Constantinople."

Lieutenant Stanfield and a section rode away to watch the Turks, while we posted outposts on the Sandhill crests around the oasis to warn us for fight or flight should a Turkish patrol endeavour to surprise the camp. Those of us not on duty gossiped and smoked under the palms, Our rifles across our knees, all keyed-up and lively and quite contented with things. The boys boiled the jack-shays as the morning wore on, cannily disguising the tell-tale wisps of smoke. Suddenly, crackling rifle-fire broke out to the north towards Katia. The guns rolled in with their song, growling thunder across the desert.

But around us was a quietness intensifying a sense of isolation! We had ridden through No Man's Land, we were in the Turks' land.

At dinner-time the firing eased down to burst out harshly and harsher again as the heavy patrols clashed.

At four o'clock our section mounted and left the oasis to join up with Lieutenant Stanfield, at the well of Bir Wazet, three miles to the south. The desert was all clumpy sand-mounds growing stunted brush affording admirable cover to man and horse, so we advanced cautiously—and halted abruptly when we spied a horseman motionless amongst the bushes. We spread out instinctively and turned towards him, but he disappeared. We stopped, craning our necks to peer above the sand-mounds. Five minutes later his head appeared like a black shadow above the bushes. We rode on and so did he. His wariness exasperated us. We broke into a trot—so did he. We cantered—He swiftly sped up an opposite ridge. We halted—so did he. We walked on—so did he.

We felt certain he was an Australian though we could only glimpse him now and then among the bushes. We were exasperated as well as curious, for he was taking us right out of our course, but we dare not ride out to Lieutenant Stanfield's hiding possy and perhaps have a Turkish patrol trailing us. So we walked our horses on and whatever we did the unknown horseman did too. He reminded me of a cunning bird, fluttering a little farther ahead only just out of reach.

We halted—I stood up in my stirrups and waved and yelled. He stopped, gazing around, but as we came on he walked off again. I reined in my horse, stood up and yelled and cursed him in lurid Australian.

That settled it. He halted, eyeing us very suspiciously as we approached, crouched in his saddle for an instant getaway.

Then we saw the rifle-muzzles of his patrol poking from the bushes.

The cunning beggar had led us right up to them.

He was a corporal of the 7th Light Horse, and was as surprised to see us as we were to see him. They were a Wandering Patrol, miles away out from the lines.

We turned away south again and soon reached Bir Wazet the tiniest of wells, hidden at the foot of two barren sandhills. How on earth did the long dead desert people dig that deep, tiny well, and what miracle told them of the presence of the cool water far down in that desolation of sands? There is no sheltering oasis, but under a dense clump of bushes was Lieutenant Stanfield, clothed in his usual grin, with his four men. The four then rode away back towards Nagid. Stanfield and we four were to remain and before morning ride out to Hill 200, which overlooks the main Turkish camp in this district.

Stanfield's section had seen numerous Turks, one patrol of a hundred men had ridden right past the bushes where they lay hidden. So our section expected some fun in the morning.

That night the four of us spent in reliefs, one man holding the horses and keeping a sharp lookout for an hour, then waking the next relief by kicking his ribs. How we blessed the thorough training the horses have had. For apart from whinnying, not one stamped its hoof. They never moved, just kept quiet and still as the sentry. Any man who snored or groaned got the sentry's boot in his belly quick and lively.

Before dawn we were plodding towards Hill 200, six miles away, invisible in the darkness. The bushes swished against our knees. We were all eyes and ears and tingling nerves. In the grey chill of dawn we distinguished the base of the hill rising massively all gigantic shadows before us, a solitary giant, looming among countless sandhills shrouded in bushes. The hill centre is divided by a deep and narrow valley a mile long which runs straight up through it to the foot of the hill peak, which was our objective outpost.

The rosy tips of the sun were rippling the sky when we rode into that dark valley, its prehistoric walls in half-guessed shapes of wind-blown sand. The only sound the squelching of the horses' hooves, no hoarse challenge, no rifle-shot. We rode right to the top of the valley where it closed in directly below the hill peak. We dismounted, leaving one man with the horses. His was a nervy job, our freedom, if not our lives, might depend on him.

Cautiously we climbed the precipitous slopes, looming far above. We crawled out on the peak edge and then lay gazing away down into the mists floating over the low hilly country. The sun's bright rays quickly dissolved the mists disclosing the Turkish camp. On the skyline about a mile away through the field-glasses we saw men moving about. Then on all the little ridges we watched the Turks turning out. The sun rose higher and we could see them distinctly without any glasses at all. It was a curious little thrill to thus gaze on Jacko in his nakedness, as it were. He was not hidden in a trench here. It felt like spying on a man's private life.

While each man of us was making discoveries I spotted on a sand spur to the north, the most advanced Turkish outpost, six men, just turning to. One chap stood gazing in our direction a long time, then another came up to him, while another standing motionless gazed towards us. We just "froze" praying they could not see us. After a while the six of them picked up their equipment and leisurely walked back down to the camp. Exactly the same thing was happening to the south of the spur, where we spied another outpost. Soon the low hills fronting us were dotted with men and about three miles away where Bir-el-Mageibra lay just hidden behind a sand-ridge, there arose spirals of smoke, from the main camp-cooking fires I suppose.

Immediately fronting us on a small ridge, a big fatigue party came marching along and set to digging trenches. We could plainly see them jumping in and out of the trenches, and the dirt flying over their shoulders as they shovelled. Scattered farther back all over the field, men seemed to be walking aimlessly about. Then on the skyline towards the smoke came laden camel caravans in silhouette. We sighed for guns, for the smoke of their fires and those lurching caravans would have made such pretty targets for shrapnel.

We watched a while longer, then took note of our own position. It was a death-trap. The approach up the valley was commanded by both sides of the hill, besides the hill-top which looked straight down the valley. The Turks had only to conceal a few swift camelmen around the opposite side of the hill, wait for us to enter the valley, and then close the mouth.

Evidently our trip to Constantinople could be a reality. Presently we climbed back down into the valley, mounted our horses and warily rode back to Wazet. At the well were Lieutenant Broughton and his four men, coming to relieve us. We explained the position and all its chances to them, and then rode back to Nagid. A telephone message along that cute ground-wire had come for Lieutenant Stanfield saying that two British staff officers were coming out that night. The lieutenant was to take them over what country it was safe to do so, as they were to spy out artillery positions. If it was quite safe they were to be taken to Hill 200 to catch a glimpse of the Turkish camp, but on no account were they to be taken into the slightest possibility of danger: their services were exceedingly valuable and their training was such that they could not be replaced.

The lieutenant received this news glumly. What a comical way the Big Heads have of looking at things.

After dusk we slipped out of the oasis into the donga so that if the Turks came in the night we wouldn't be there, divided the horses into four parties so that each man on duty could watch a fourth sector of the skyline around us. Then cuddling our rifles, we curled up on the sand and went to sleep.

Men of the 2nd Light Horse in action behind the stone barricades at Nalin, Palestine.

18

July 1916—Before daylight, we warily returned to the oasis and boiled our jack-shays. How appetizing the bully-beef and biscuit tasted! It was late (how very lucky for us) when we mounted and rode out, Lieutenant Stanfield, the two artillery staff officers, and five plain Aussies. The staff officers we soon felt friendly towards. They found everything interesting, and were plainly expecting an exciting trip. Their bobtailed neddies were groomed to the last polish. They were smartly dressed themselves, a contrast to us unshaven, sunburned bushrangers in breeches and flannel, rifle and bandolier. No doubt if the Turks captured us the Constantinople papers would issue triumphant articles describing how raggedly clothed and poorly equipped the Australian troops are, exactly as we have erroneously supposed some of their regiments to be from similar prisoners we have taken.

We passed by the small black mouth of that lonely well at Wazet and soon afterwards spied Lieutenant Broughton and his patrol riding towards us through the bushes, coming from Hill 200. It had been their lucky morning. Broughton had been naturally suspicious as to the safety of the Observation Post, so instead of going up the valley he had circled away searching for a new approach, had crawled up in the dark and had heard figures moving around the Post. Naturally they had not waited. Jacko must have been annoyed when the dawn did not bring the birds into the trap.

The patrols then separated, we going ahead very cautiously, the staff officers all smiles, the sun shining on the fat rumps of their horses.

When a mile from Hill 200 we saw two figures silhouetted on the skyline gazing in our direction. But we had approached under perfect cover and almost certainly they had not seen us. We rode on and the figures vanished, but shortly afterwards, a group of eight appeared. We manoeuvred accordingly ever closer towards the great hill for we had to find out whether that portion of the Turkish army behind it were marching out to cut in behind Romani; I grinned as I glanced at Stanfield and remembered the two staff officers.

When half a mile from the foot of the hill, we discovered that a ridge on the left side of the valley was lined with men. We halted, laughing, the staff officers loudest of all. Simple old Johnny the Turk: it's a wonder he didn't come right out and try to put salt on our tails!

So we measured the chances, our eyes and glasses so carefully spying the long sloping right-side hill of the valley. There were no bushes on that side, it was all golden sand; there was no cover there for an ambush.

Excitedly one of our chaps distinguished a hidden outpost on the left side of the valley, much closer to us. We spread out in exhilarated mood and cautiously edged towards the mouth of the valley. Instead of riding into it, Lieutenant Stanfield slewed around to the right and presently we were climbing up the mile-long slope of the hill. Now right up that hill, forming the "lip" of the valley is piled one of those characteristic "razor-backs" that are churned up by the desert wind. This razor-back was simply a mile-long rampart of sand, from seven to ten feet high and twenty thick, running right up along the valley top. We clung to the hillside edge of the razor-back and it thus screened us as a wall would from the Turks on the opposite side of the valley.

When half-way up the hill, we stopped by a depression In the "wall." From here, still sitting on our horses, we could peer over the razor-back across the valley. What exclamations as we spied the whole parallel side of the valley lined with men lying on their bellies. How we laughed. It looked absurd, almost a whole regiment had been turned out to catch one tiny patrol. But, by Jove, how alert we felt!

Then away up at the top of the hill where our Observation Post was, a camel sauntered across the skyline. We grinned, expecting a shot every second. None came. The Turks were obviously nonplussed by us not riding up the valley. They lay there with their rifles at the "ready," waiting for our next move. Presently away up by the Observation Post, a cavalryman rode out into full view just as if he hadn't seen us, and started to urge his horse down into the valley.

Such a barefaced decoy! I thought we would have had hysterics as we watched the cavalryman come within an ace of breaking his neck when clambering down into the valley. How he must have cursed us in all the names of Mohamed! Then Stanfield said: "The game is up, boys, but I'd like just one glance down behind the hill. There may be a movement of troops."

So the two artillery officers and two men stayed there looking across at the Turks and keeping an eye out for camelmen that might make a dash from somewhere around the hill, while Lieutenant Stanfield, Sergeant Paul, my temporary section mate and myself made all speed up the hill, watching out that our horses did not plunge into those cursed "sand holes" full of floury sand in which there is no footing.

We quickly reached the hill-top, but reined in just where the end of the razor-back curls around to our old Observation Post. I jumped off my horse and started clambering up the razor-back with the horse's rein over my arm. Stanfield and Paul jumped off and handed their horses to the fourth man, Stanfield taking his rifle. We peeped gingerly over the razor-back and there a few yards away were Turks lying down all around the Post, their rifles at the "ready." Bedouins crouched amongst the bushes. All down the left side of the valley were waiting men. The sight of all those waiting Turks took our breath away. Three big chaps crouched up trying to level their rifles through the bushes to get a clear shot at our heads. I jerked up my rifle, but at that very moment rising opposite us from the other side of the razorback only twenty feet away appeared the black cloth elbow, then the rifle muzzle, and then, very slowly and cautiously the head of a Bedouin. I bit my lip, in taking aim, to keep steady, then crack, crack, crack, three bullets in a second and the Bedouin clenched out his arms and bit with his mouth in the sand.

"On your horses, boys; quick !" shouted Stanfield. One jump and we were down the razor-back. "Steady, boys, there's plenty of time," shouted Stanfield; then immediately each man was in the saddle he dug the spurs in and laughed as his horse leapt away, "Come on, boys, and we'll give these——a go for it! Ride like the hammers of hell!"

In a bound the four horses were away and in the same breath we could see the sand spurting around the artillery officers and men half-way down the hill, then in one dreadful second my horse stumbled—and fell! My Heavens, he rolled over on top of me—I thought that if I lost my head in the next few seconds I was done. I just heard Stanfield's vanishing shout: "Whose horse is down?" while I snatched at the struggling beast's head as he floundered for a foothold, madly excited. I snatched the ringed bit and held him steady: he quietened instantly as Sergeant Paul galloped back. My head was buzzing, but steadied like the horse when Paul said quietly: "Steady, lad, there's plenty of time. Mount him again." I let the rein loose and the horse made a desperate effort—I snatched the rein as he floundered up and he pulled me up with him. I dared not raise my eyes to that razor-back. I understood now that everything happened so quickly that the dead Bedouin's mates dare not raise their heads, not knowing what was awaiting them. I swung into the saddle as the first bullets hissed by and in seconds we were galloping through what sounded like a shower of red-hot hail. How our horses raced to rejoin their mates galloping far down the hill! I had one foot in the stirrup, the other stirrup-iron and the feed-bag were doubled up under the side the horse had fallen, but I could feel by the plunging gallop that the neddy was not going to fall again, and that ride was simply grand!

Stanfield and the Aussie were rapidly gaining on the lower party who now were galloping almost at the bottom of the hill, the sand spurting up around the flying hooves. Paul and I rode like laughing madmen—we expected any moment a horde of camelmen to come from the low country and cut us off.

As we neared the end of the hill the razor-back was much lower and half our bodies were in view of the Turks. How their rifles rattled from across the valley! They must have enjoyed those few moments. Quickly we gained on the rest of the patrol, the wind swished back their excited laughter. I glimpsed one chap's face with his eyes staring from his head in excitement. A second later and a bullet flipped off his hat. At fourteen hundred yards' range I noticed a peculiar thing; I could not hear a single rifle-shot, only the *zip, zip, zip, zip* as the bullets sped softly past. I think the wind rushing past our ears blocked out every other sound.

Soon after the others had sped out into the desert from the bottom of the hill we caught up to them, steadying down to a swift hand canter. Everyone was laughing, the staff officers were enjoying the joke immensely, their horses exulting in the gallop. "By Jove! Hear those bullets whistling. Better than being in India, old bean, eh what!" The little excitement quite broke up their English reserve. They were cantering beside the two long, hard faced horsemen who had stayed with them, and the Aussies were convulsed by the English terms that seemed strange to our ears. "What an experience to tell the chappies! We'll be the envy of the mess. Hear that little canary, I believe it's lost its jacket. My hat!" And so on as we sped through the bushes into the distance. But I was as excited as they. I lived through that ride.

19

At Nagid that night, we were keenly expecting Jacko to come after us. Next day, the spying section climbed a distant spur of Hill 200 and quietly watched an outpost and thirty camelmen who were lying in wait for them. As the day wore on we in Nagid got tired of listening to the gunfire, so went out and pleasurably indulged in chasing Turkish patrols from Hill 386. But next day they returned in force and chased us. We outwitted them though. One man got a shock and a great bruise over the heart—a bullet flattened itself in his bandolier. The Turkish forces soon commenced to advance all along their line, which news Lieutenant Stanfield phoned to headquarters. Every hour we expected to be run to earth at Nagid, as our hidden posts spied distant bodies of troops pressing forward towards Romani. Then we detected their advance patrols marching towards Hill 426, which left little us isolated in their advancing centre.

The rolling of rifle-fire and booming of guns was ceaseless now, day and night. The afternoon before we were withdrawn, Lieutenant Stanfield's party captured a Turkish rifleman hidden near our well at Wazet. Next morning, the day on which a party of New Zealanders was to relieve us, Lieutenant Broughton with his four men brought in four Turkish infantrymen who had been hidden by the well. The En Zed relief arrived shortly afterwards. They told us there had been constant fighting ever since we left Dueidar; but what interested us locally was that at Hamisah, only two miles to the north, last night all of a Light Horse standing patrol had been surprised and killed. So the Turks had drawn in right around us. I don't think the En Zed patrol will be able to stay at Nagid for long.

So that's that. Here I am back at the old possy (thanks to Sergeant Paul), and have written up the old diary 'tween duties. It was very game of Paul that day, to wheel around and come back for me. He need not have done so, because each man knew we were on a duty in which at any moment it might be a case of every man for himself.

The taubes gave Romani a terrific bombing this morning.

…The troops have been fighting day and night ever since we left, in fierce charges—attack and counterattack—galloping their horses to within point-blank range of the Turks—scrambling on their horses when the Turks came raging around them, only to gallop back to turn around and fight again. Men and horses are desperately in need of sleep. The guns are boom-oomoomming.

August 1st — Poor tucker again. The Turks are nearly at Katia. They are swarming towards Romani … More taube alarms. Desperate night-fighting out in front — some of our night outposts have been bayoneted.

From Dueidar far out fronting Romani right to the sea coast we have a line of Listening Posts and Cossack Posts. These posts ride out every night — wee bunches of men creeping close to the Turks. The Listening Posts listen! If they hear the Turks coming they fade back into the night without firing a shot. The Cossack Posts, on sight or sound of Turks, open a furious fire — leap on their horses and gallop back for their lives. Both systems of posts are to warn the main body of each successive Turkish advance.

Nightly, the Turks advance a few miles across the desert, thousands of men in far-flung waves — the first wave sheer desert fighters and as noiseless — they seem to arise out of the very sand in overwhelming numbers — some of our Cossack posts have been bayoneted as the men were springing on their horses. Under the brilliant starlight, the Turks when arising from among the bushes appear like transparent shadow men, the harder to realize as our men are dazed for want of sleep. And yet, two mounted Australian brigades and a New Zealand regiment have delayed a Turkish army for more than a fortnight!

August 2nd — Heavy gun-fire. The rifles rattle in waves of sound. Nightly these waves come sighing across the desert, then they trail away to a moan to crackle out again with a rattle that swells to a roar. We can distinguish the attacks and counter-attacks just by listening to these receding and roaring waves of sound.

August 3rd — The 1st Light Horse Brigade with the little Tommy guns in a galloping charge cleared the Turk right out of Hod um Ugba. But he fought back the 2nd Brigade from Oghratina to Mageibra. His strong-posts are now within a mile of Katia and his camelmen have forced their way into Hamisah to the south. The Big Scrap draws very close … Our patrols send word that two thousand Jackos have appeared away out west of Bir Wazet.

August 4th — Last night, six men including myself had the wretched luck to be sent to Hill 383 for night patrol. At twelve o'clock a message came through the field-telephone that the regiment was moving out to delay the Turkish left flank now advancing on Nagid. The En Zed Scouting patrol that relieved us phoned that the Turks are advancing from Mageibra in great force. So our Heads anticipation of this move on the part of Jacko was correct. I wonder how the little En Zed scouting patrol has fared!

…At three o'clock this morning heavy machine-gun fire broke out about two miles to our front. It is terrific now.

…A New Zealand regiment took the place of the 5th at Dueidar. They sent out a Patrol to relieve us—we are now back in camp, all ears to the boom of guns just out in front. It is much the heaviest desert firing we have heard yet, the Turkish waves are breaking in on Romani. We are mad at being isolated from the regiment at such a time.

There were big bayonet charges at Romani last night. The Australians suddenly sprang to it as thousands of throats yelled "Allah! Allah! Allah! Allah!" "Finish Australia! Finish Australia!" and the Turks were right on their main line. Cossack Posts out in front were bayoneted in seconds. There is bitter fighting now on Mt Meredith, Katib Gannit, Hod el Enna, and Wellington Ridge. The Australians are slowly falling back. We're only getting scraps of news.

…The New Zealand regiment is moving out. Their brigade is going to smash the Turkish left flank before the Australian brigades are overwhelmed. These have been fighting day and night this fortnight past—their Casualties are very heavy. We saddled up to ride ahead of the En Zeds and try and rejoin the regiment, but the Scotch colonel in charge of Dueidar stopped us. He's got a damned cheek; we are not Scotchmen.

8.30 a.m.—A regiment of English Yeomanry have arrived—they are sending their screen out at the gallop: their regiment is now at the trot— we wish them speed to help the Aussies. The guns are booming close and angry. I wonder who is winning?

…We have just heard that the En Zed patrol at Nagid has been captured! Caught in the night between two columns of infantry. What a hell of a time they'll have trudging back across the desert to a Turkish prison camp hundreds of miles away! Poor chaps! I'm glad it wasn't us.

…At last we are to be allowed out. The Turks have cut the telephone-wire: we are to go and repair it, after which we can go to hell, so far as the Scotch colonel cares! He swears the six of us are more damn worry than a whole battalion of "soldiers!"

9.15 a.m.—Hell and blasted Tommy! We were just mounting when an orderly rushed over from the C.O.'s tent yelling that the telephone-wire is now in enemy territory! They must be making a rapid push. The guns are banging snappily; the machine-guns have the throaty chorus that means business close by … Taubes are threatening Dueidar. Lieutenant Hicks's groom has just returned. He tells us the regiment met the Turkish left flank advancing in thousands. The 5th held them up for several hours but were forced to retire at the gallop. Lieutenant Hicks is wounded.

10.30 a.m.—'Planes are fighting overhead, rifle-fire is rattling far across the desert—how vicious it sounds at times! Austrian howitzers are roaring.

3 p.m.—Numbers of the Yeomanry who hurried out this morning are back again wounded, riderless horses keep galloping back to Dueidar.

…Mounted troops are charging for miles along the Romani battle-front; the Turkish waves have reached almost to the redoubts. Fighting has been severe all night and apparently very uncertain. Captured orders say Von Kressenstein is the German general in charge of the Turkish army. His orders are to take the Canal at all costs. Well, he has struck a small but trained Australian army. General Chauvel has his work cut out though—by the sound of it.

4 p.m.—More cursed taubes have sent us scattering over the desert. It is amusing to see the Gyppo camel-drivers running for their lives—they can go some too—when those nasty taubes buzz low overhead.

5.30 p.m.—Hoo-blooming-hurrah! The old regiment comes riding back. We feel like naughty children who have been smacked and put away in a room.

The regiment had a very lively time holding the Turkish reinforcements up. The main Turkish columns are now hammering all along the Romani front except at the infantry redoubts. The 1st Light Horse Brigade has lost most of its officers. News of all sorts is trickling in like the Yeomanry horses.

7.30 p.m.—Brigadier-General Antill with the 3rd Light Horse Brigade has arrived. Word comes that the Turkish attack on Romani is broken—the New Zealanders and Yeomanry crumpled their left flank—they are retiring to El Katia. It was a very close thing; some of their battalions actually got to the railway line built behind Romani. Thousands of the Turks crept up the precipitous sand ravines between the ridges. The Australians in the early fighting hours could only distinguish them as shadow men. Both sides blazed away point-blank at one another's rifle-flashes, which when the quarter moon had set made the intense blackness seem aflame. Their main attacks captured Mt Meredith, Wellington Ridge and apparently other positions … The guns are booming sullenly. The regiment is ready to go out again at a moment's notice.

August 5th—Before dawn. We are moving out.

20

Unbelievable that all could happen in only a few days! Here's to the diary while my mind is fresh to record just a little of all that excitement, endeavour, interest and tragedy.

We took the El Katia track. Dawn came with a crimson sky. From 383 ridge we gazed behind at a grand sight all lit up in pink and grey and khaki stretching right back past the redoubts of Dueidar, a winding column of New Zealand and Australian mounted troops.

We laughed with the exhilaration that disciplined comradeship brings, then rode on towards the boom of guns, harsh now and sinister. The sun blazed at Bir-el-Nuss where we watered the horses, and waited through a hard expectant two hours while up rolled more Australians, more New Zealanders, and finally the helmeted Yeomanry. And then came the Somerset and Leicestershire Batteries R.H.A., all chirpy and spoiling for fight, then spare battery horses prancing and fat. Then long lines of ambulances ominously rolling along. So regiment after regiment, brigade after brigade watered their horses. Others took their places, while still others were riding up behind until the oasis was surrounded by a dark brown cloth of horses and men. But the faces of the men who had been fighting day and night were haggard, their eyes starey, their horses very tired.

We moved off again, whistling and singing, laughing and joking, the horses of us fresh brigades pulling at their bits, seemingly careless of the fierce desert heat.

All in our shirt-sleeves except the Yeomanry. all hardy, laughing fighting men. And in the centre, the Tommy batteries with their caterpillar wheels and well-conditioned, fresh-faced little men, and the guns that soon again were to back us up so well. It was grand to feel that a man was one of them all.

As we rode along, we heard scraps of the last twenty-four hours' fighting. Of the shadows that arose from the night and annihilated the Cossack Posts—the Tasmanian sergeants of two posts of sixteen men were the last to be downed (and they died still fighting from the ground). Of wild starlit gallops right through the Turks. Of an officer who wheeled back into a horde of Turks when four of his men's horses were bayoneted two men leapt on the horse, one clung to each stirrup and the five actually got away. Of a big Queenslander, too—the troops are still laughing at his rage. In the shadowy light he thought one of his men was down, he wheeled roaring into the Turks—he lifted the man straight up on to his saddle and galloped back unharmed to find that his rescue was a Turk!

There are sadder tales too, of men beaten in the fighting retirements by the heavy sand filling their leggings and boots, plodding desperately on only to be overtaken and bayoneted. Of numerous Turks, too, so eager in the attacks that they threw away their boots so that they could press over the sands faster. Of the terrible scream when eight thousand Turkish bayonets glinted in the starlight and charged Mt. Meredith. Of the "shadow shooting" as the "Men of Allah" leapt up from the Romani bushes. On the precipitous slopes to the south a handful of men peering over the brink shot the Turks down like wallabies and they rolled over and over and over far down the walls of sand. The slopes were like long golden sheets faintly alight, the Turks climbing up distinct as coal-black shadows. Of Chauvel, who at dawn in person came to the rescue, trotting into action with the 2nd Light Horse Brigade from Etmaler. We heard too, only in short sentences, of the heart-choking struggles to get the wounded away almost from the feet of the oncoming Turks. A Corporal Curran of the 7th brought seven men to safety before he was shot dead. Of the little Tommy batteries, quick and willing, ready for action anywhere, everywhere, wiping out machine-gun teams, helping the horsemen again and again immediately the dawn came.

And those who would talk, every man in his own fashion, by praise or jest or grim curse, expressed admiration for the willingness, the determination, the bitter stubbornness of the Turk. The boys marvel at the breathless escapes of Brigadier Royston—"Galloping Jack" they call him. He wore out fourteen horses, galloping up and down the line to wherever the fight blazed the thickest. We heard of the last bayonet charge of the Wellingtons and 7th Light Horse when they charged Wellington Ridge and two hundred of the Turks would not throw down their arms but stood up to the steel. After it was over, Colonel Onslow and three troopers advancing again were shot by snipers only yards away as if it was only the one rifle-shot—the four men fell as one.

Through it all we can sense a battle very, very narrowly won. As the columns rode along it seemed to me that these men with their strained, grimy faces, their eyes fever-bright, were living on the terrible excitement of hours before. Some were swaying in their saddles, their heads, like their horses' heads, drooping for want of sleep. And we were riding out with these overtired men to deal another blow at a furious enemy at bay!

As we drew level with Romani the roll of the guns to our right grew like thunder. The day was fearfully hot, the sand heavy. We branched off towards El Katia, chewing biscuits as we rode. Then my neddy snorted at a Turk who lay with his sand-filled eyes glazed to the sky; then past a smashed case with its Turkish ammunition shining on the sand. Then the neddy must snort at a dozen Turkish stretchers, all blood-soaked, then the guns boomed menacingly as we rode past a quiet oasis strewn with hundreds of ammunition cases, Turkish stretchers everywhere, odd ones occupied by dead men. We joked at the numerous bottles—mute evidence that the Turk and German officers had drunk to "the day."

A sharp burst of rifle-fire rolled out into a clattering roar. We were drawing very close! That old, peculiar feeling stole over me, it always does when going into action. I wonder if my ancestors experienced it when they advanced with club and spear.

"Halt!" "Taube! Taube! Taube!" was shouted back along the line and all those columns of men and horses stood perfectly still while the bird of ill omen droned by overhead. What a miraculous target she missed. So much for the spotting-powers of the aeroplane! We pushed on and the rut-tut-tut purr-purrrr of machine-guns grew ever more plainly amongst the rifle fire. Then crash! crash! crash! crash! and in the sky almost overhead circled our own 'planes with Turkish shrapnel exploding all around them. Presently we halted while the Heads held a pow-wow on the ridge in front. The firing, sharp and clear, sounded just over that ridge. We dismounted for a few moments while the following troops massed behind. Then mounted and as a massed regiment rode right to the top of the ridge. Straight down in front of us was a mile-long slope of hard sand leading right to the first oasis of the great El Katia system.

Shrapnel was bursting over all those miles of palms, snowy puff-clouds distinct above the dark green of the oases.

The colonel pointed to the oasis immediately in front "A battery of heavy Austrian guns has been located in that oasis," he said quietly but distinctly. "We have to charge and take the guns. Regiment—fix—bay'nets!" By Jove, what a thrill ran through the regiment! The flash of steel, innumerable click, click, clickings and five hundred bayonets gleamed. As the squadrons moved out over the ridge we broke into a brisk trot.

The colonel with the adjutant rode in the lead, the squadron majors leading their squadrons, the old doctor and padre riding knee to knee, laughing as if at a great joke. The oasis was coming nearer and nearer. We held the horses in so as to have their strength in the last great crash. But they were getting excited; the men were getting excited; we rode knee to knee; to right and left were excited faces and flashing steel; and our bodies felt the massed heat of the horses that tugged amid strained as the squadrons broke into a swift canter. Then a horse reared high as it screamed and we went into a mad gallop, the horses' mouths open and their great eyes staring as the squadrons thundered on.

The officers waved revolvers and shouted, a roar of voices rose up and in the mad laughter was shouted wild things and the oasis just rushed towards us. Then to our right I glimpsed a sandbag trench. "The right flank is gone," I thought, and stared straight ahead with gritted teeth—but not a solitary machine-gun rose above the parapet as we thundered by. I breathed again, thrilled, as hesitant men lined the oasis edge—even in those last mad moments I sensed that they were too terrified either to run or fire. We crashed through them and thundered into the oasis to a wild crackling of palm branches—we crashed right through the oasis out into a plain surrounded by a mile of palms. And there was the colonel out in front with his horse pulled back on its haunches, his hand held high. "Halt!"

We surged around him in hundreds on plunging horses we gazed at one another all jammed up—the guns were not there! We breathed—then came the sickening thud of bullets into horse-flesh. We doubled back amongst the palms with a machine-gun's bullets spraying into us. But what luck! other guns were firing too high and raining down dates upon us from the palms!

The colonel's horse was down. Panting horses with crimson chests were sobbing up against their fellows. Back among the palms we got shelter and quickly lined up in troops.

Captured Turks told us that before we charged they hurried their guns back to the larger oasis and surrounded that central flat with machine-guns. If we had charged across it I don't think a man would have been left alive—it was a morass in the centre!

"Dis-mount!" We handed our horses over and Morry tried wildly to swap his place with any other man who was going into the firing-line. But no one wanted to be a horse-holder. We pushed out in an encircling movement to the right, following the clumps of palm-trees. In among these were mounds of sand all thick with stunted bushes. As we worked thus towards the big oasis the Turks sent their shells crashing down among the palms: our little Tommy guns firing at close range from the open desert replied as if they would belch their very throats out, but they had no hope against the heavier shells amid much more plentiful guns of the Turks. Machine-gun bullets chipped lathes off the trees and rifle—bullets squirted the sand.

As we moved off, a New Zealand squadron came galloping into the palms—the Turkish machine-gunners had the range by then and the horses screamed down with the riders crashing as their outflung rifles thumped the palm-butts. Then right away to our left we saw dismounted men, wearing helmets, taking every advantage of cover, running, firing, lying down, jumping up to run forward and lie down again and fire until they spread out on the edge of the flat, to pant on and suddenly flounder in morass I don't know what regiment the poor chaps were, but they never crossed the flat.

The heat was fearful. Rifles were almost too hot to hold after the first twenty shots. By Jove, I was thirsty! We pushed on and the palms echoed back the rifle-fire and shrilled with bullets. Shells from our own guns streamed overhead to crash upon the big oasis in front. A few New Zealanders panted up near us—they kept in a scattered group, running and firing, running to farther sand-mounds to drop to their knees and blaze away again the sweat streaming from their red faces and bared chests. With a furious resentment we saw that they had located snipers (we hate snipers) in a clump of hushes ahead of us—bullets whistled spitefully close to Stan and Bert, but it was so dreadfully hot that we just crouched ourselves up and trudged on. Suddenly the New Zealanders all jumped up and raced forward firing. Immediately from a bush a white rag waved and from each of score of bushes up stood a Turk with his arms raised. It would have served them jolly well right if they had been bayoneted. A sniper just kills as many men as he can from ambush, then when surrounded simply holds up his hands. However, he is a game man to take the risk.

A Look-out Post of the 3rd Light Horse at Khur-betha-Ibn-Harith.

21

Other brigades came pounding into the great oasis. But we could see only occasional glimpses of struggling men in groups among the palms and sand hillocks. I felt sorry for the 1st and 2nd Brigades; three days and nights without sleep and now under hellish fire, struggling through the swamp away out on the left. The Turk was fighting in a snarling fury over every yard of ground. There was no trench work here. You were all concentrated just in your section and often for mad, hysterical moments, just in yourself. You saw a Turk's head in a bush, you saw his moustache, you saw his eye glaring along his rifle-sights. You fired too, with your breath in your belly, then rushed forward screaming to bayonet him, to club him, to fall on him and tear his throat out and he met you, a replica of the berserk, frightened demon that was in yourself.

We pressed forward gasping in the frightful heat. I swear the last drink from my water-bottle was boiling-hot. From palm to palm, from mound to mound, we fought forward, sweat creasing rivulets of sand down our faces, matting the hair on bared chests, dirty brown blotches on our arms bared to the shoulder.

Then, fronting poor old A Squadron, gleamed a long, bare patch of sand, like a road running through the Lushes. We held our breath—I hesitated; I cannot blame myself—in burning fear I plunged forward. Hell! they waited for all the desperate squadron to get far in that open patch. My God! not even in Lone Pine have I heard such a hellish concentration of rifle-fire. My legs were lead, the yards seemed miles. It is a terrible nightmare when a man strains his very heart out and cannot move an inch. The air was just one sizzling hiss—flying bullets at point-blank range possess an awful sound. At. the bushes edge I flung myself against a sand-clump in frantic fear. The poor fellow running beside me just reached cover when he spun around and around desperately hit. It must be agonizing, the shock of a bullet tearing through a man's stomach just when he is panting for breath.

It took us a terrible time to get our breath enough to be able to push on again and sight the rifle steadily, beating back the Turk from bush to bush. What was happening elsewhere we could not see and had no time to look. We could see Turks and they could see us, and we dare not shift our gaze for a second. Captain Bolingbroke led us on splendidly. The other squadrons spoke enthusiastically of their officers after the fight was over. The little bare patches of sand, thank God there were no more like the big patch, were hell to cross. Over every bare patch the Turks had machine-guns trained—as the men in straining groups dashed forward they were not fired at individually, they simply had to rush through a continuous stream of bullets—over some patches the machine-guns were so trained that several streams of bullets crisscrossed. It was in crossing these bare patches that most of our men got hit.

So we pressed slowly, but surely forward, ever nearer the huge Katia oasis, and the firing flamed to a roar, then to a point-blank crackling that rebounded among the trees. The mystery was that a man could live through it. The cover saved us and the simple fact that we were shooting the Turks faster than they shot us. Captured orders have proved that since. Suddenly Bert shouted in a laughing, high-pitched voice: "A well! Boys, here's water!" All near enough to hear him rushed over. Damn the Turks! There were a score of half-filled Turkish water-bottles lying dropped by the well as we rushed up. The water was ice-cold—I filled my bottle and drained it—other men came panting up—we clustered like bees around the lip of the well. I don't know if it was the Holy Family Well famous in that oasis, but its water was sheer heaven.

The Turks saw the target. A bullet smashed a man's temple and his blood sprayed the clear water surface. We pulled him out and thrust down our bottles again. As we drank and filled them again we would rush to the encircling bushes and shoot the Turks back while other men came panting up for a drink. Quickly we formed a determined post around that well and I don't think all the Turks out of Hades could have taken the water from us. Bolingbroke let the squadron have a thorough drink and get their wind back, then we went on again, laughing, but grimly ready for the Turk.

Soon we got into quite a hot spot—a series of stumpy sand-mounds each with its clumps of shrubby bush. Turks waited behind every mound—they fired "on the blind" through the bushes. We fired back, split up into many little groups, nearly all in sections. We'd jab our bayonets through the bushes, and the Turks would stab back—we'd burst in around the bushes and glimpse the Turks' gasping mouths as they hopped back behind the next mound. Sometimes a section would get its Turks; we always knew when by the yell. The Turks are great at that type of fighting, but by Jove, we beat them!

We fought right through that series of mounds and then the oasis spread out more open again. I got a glimpse through the palms of the big circular flat, amongst the bushes encircling it I could see little bent-up men running; they reminded me of wallabies going for their lives. The smoke from the bursting shells was like a white vapour rising over the flat. Away back where we left our horses, heavy shrapnel-clouds were floating. I noticed stray neddies galloping through the palms and hoped our section horses were not hit. Then a team of German machine-gunners opened out on us and I had no time to look around.

We fought on and gradually pushed the Germans back. We tried hard to take their guns, but they were too well supported by infantry. The squadron had lost heavily by now and as men got wounded other men left us to take them away. But they always came panting back when they got the wounded under partial cover. We have a sort of unspoken oath always to get our wounded away.

The afternoon was waning rapidly. We wondered what was going to happen. The Turks, in some cases anyway, were decent to our wounded. Our chaps had to carry them over those bare sand patches, and of course they could not run. But twice through the afternoon I noticed little knots of men stumbling across the bare places with wounded, and the Turks did not fire. But when the men ran back to rejoin us the hail of machine-gun bullets simply whipped the sand into dust.

Yard by yard we drove the Turks back into the very heart of their last big oasis. But too late the sun was setting. The big oasis was the hub of streams of machine-gun bullets; all our troops seemed to be fighting now in isolated squadrons; we needed far more men than we had.

Captain Bolingbroke started back alone for orders. In the first clear patch a machine-gun spat bullets around him, he flung himself flat on his face, rose again and crawled forward. But the machine-gunners evidently could just see him and the way the captain ducked was laughable. We all laughed; long and gaunt, he looked so comical. He winked back over his shoulder.

Cunningly he rose to his knees, but flattened instantly under a spray of bullets. He crawled the rest of the way to the bushes on his belly. It must have seemed an eternity with his lips breathing against the scorching sand. A man is lucky in this war not to a have a tail.

We fought on, waiting for the captain to find out what was doing. One poor chap was shot dead beside me. He slipped over without the slightest sound. He was the man who took Morry's place in the section. Stan and I hurriedly scratched a hole in the sand. There was only cover behind our mound for two men and the bullets were whipping around us. To have pushed him away meant a certain bullet through us. We dare not use him as a sandbag for he would attract bullets and we had no trench to lie in. We pushed him into the hole, whipped the sand over him, but left his face uncovered. Then we lay in partial safety across him and went on firing.

Every moment we expected the order for the final bayonet charge. It would have been simple hell had it come, though I believe what was left of us all would have taken the remnant of the Turkish army. The order to retire came instead.

I do not believe the Turks realized for some time that we really were retiring. When they did—They swarmed up behind us among the bushes. In little groups we would run back, fall behind cover, and dead-beat, wearily raise ourselves to fire at the Turks while other little groups staggered past us and fell into cover behind. The Turks were so excited that numbers of them stood up as they blazed away and gave us splendid targets. But the sun setting behind us made us a grand target for them.

At long last we struggled back to a final sand-rise over which in the gathering gloom there shot ribbons of fire as our machine-guns fired over us into the Turks. The enemy must be held back from here while the troops mounted.

That last rise was simply awful. It seemed such a heartbreaking distance away. Stan and I gathered our breath for the last run, looked at one another, then started off. The sand spurted around us, the air was alive with *zip zip zip sh sh shshshsh*—awful! I felt I was running at terrific speed, but not covering a yard of ground. We passed Bert. He was walking! I gazed at him and tried to shout but only whispered hoarsely: "My God, Bert. Run—run—it is the last!" but Bert's mouth was open—he shook his head and plodded on. Bert is short and poddy. We flung ourselves over the rise and rolled helplessly. Captain Bolingbroke was hit. I clawed myself up to watch for Bert. I watched a miracle. He just walked slowly up that rise and it seemed that all the Turkish army was firing at him. He collapsed when he got over the rise. Hours afterwards, he swore like anything.

We had splendid cover now; there was plenty of ammunition. We filled our belts and bandoliers when we could move, then fired from behind the ridge as the rest of the regiment staggered in. The doctor and padre were busy with the wounded behind the rise. Ambulance carts were hurriedly taking them away. But the Turks pressed us so close that some of us were even hit while firing from the cover of the rise. An En Zed doctor-captain and his sergeant were killed when most gamely trying to bring the wounded in.

Eventually the horses came pounding along. What a mix up in the fast gathering dark! Men shouting for their mates, for their horses; fresh men getting wounded and their mates shouting for help to lift them on to a horse. The horses excited and plunging, men leaping on, laughing as they reached to help a scrambling comrade whose horse was rearing—a burst of Turkish yelling "Allah! Allah! Allah!" as they came pouring up the rise—a thundering of hooves, a jumble of horses as we plunged off through the palms.

Captain Patrick took charge of A Squadron. He proved a good man in the subsequent fighting. Major Wright as well as Bolingbroke was hit and Captains McNeill, Chatham, and Plant. Lieutenant Waite was knocked; I don't know what others and N.C.O.'s and men. The 6th Regiment lost their colonel and lots of others.

So the Turks held Katia again and all its wells, and we had to ride right back to Romani for water. What a weary, weary ride that was! Such a few miles and so many hours to do it. Our little battery (no one knew what was happening to all the other little batteries scattered over the desert) would get clogged in the sand until the straining horses could hardly stand. The 2nd and 3rd Light Horse had been for the last sixty hours constantly in the saddle and fighting. Their horses had gone sixty hours without a drink. Our column was constantly being halted by the infantry outposts stationed around the Romani redoubts. Fighting day and night had been going on for weeks and all troops were nervy. It was nothing but "Halt! Halt! Halt!" shouted down through the darkness. We would tumble off and fall sound asleep in the sand, but only for a minute. "Mount!" and we would wake to find ourselves under the horses' hooves as they struggled to rise. Then on again through the night until just before sunrise we rolled out of our saddles at Romani and fell fast asleep on the sand.

Light Horse riding back along the ancient caravan road after a surprise attack against the Turkish force at El Mazar.

22

August 6th—At dawn we hurriedly watered the horses, put the feed-bags on, then opened some bully-beef for ourselves. But the order came "Mount!" so we rode away, chewing biscuits, back to El Katia. It was sad. Many of the sections had vacant places, hurriedly made up to strength again from other casualtied sections. So each troop rode short of men and horses. A man would turn to speak to his mate, only to shut his mouth quickly, on remembering a new man rode there.

A winding column of men was already riding towards El Katia. The sun arose blazing hot. Everyone was listening. Presently we met scattered Tommy infantry stumbling towards El Katia. Many were already crazed with thirst—we prevented numbers of them from doing dreadful things. Some were scratching in the saltpans for water, odd ones actually were already vomiting through sunstroke—some had thrown off their equipment and ran screaming and laughing across the blazing sands. Many of our chaps gave them water which they could ill spare. Still not a shot was fired!

The big dark oasis that had been alive with the sound of death yesterday was now silent as the grave. Here and there we rode past a dead horse, sometimes a forlorn group of Turkish dead, and here two New Zealanders. The oasis itself would be thick with the dead. Our own regiment had buried most of its dead as it fought.

Not a shot! we actually rode past the great oasis. The Turks had abandoned it during the night. Our infantry, straggling far behind, took possession. How they must have rushed the wells!

We rode on and a mile farther ahead saw the long screen of Australian horsemen right on the heels of the Turks. We wondered if they were the 3rd Brigade who had been held up at Hamisah yesterday afternoon. The country ahead was all low sandhills. Often shimmering. In the distance loomed larger hills. Behind us amongst the sand-dunes were the huge green carpets of the El Katia oasis system with the frowning Romani hills dimming behind.

This morning, the regiment was escort for our brigade artillery. Suddenly crack! crack! crack! crack! and the scouts away ahead were in touch with the enemy. Then came isolated rifle-shots a mile away, perhaps from the deadly groups of snipers the Turks are so fond of placing out in front or behind them. We pricked up our ears. The morning wore on. We moved on to Um-Ugba, remembered by the Wellingtons for their bayonet fight there.

There came heavy rifle-fire away out to the right where the 3rd Brigade was trying to push in the Turkish left flank by Hod Abu Darem. A roar of rifle and machine-gun fire crashed out directly in front where the New Zealanders were hammering the Turkish rear-guard at Oghratina. We were halted, and then came the waiting that is so terrible in warfare. Odd bullets whistled overhead, some plonked in the sand close by, but no one was hit.

Biscuit and bully-beef for dinner. We lay in the shade cast by our horses, some smoking, some talking, some sleeping. We felt sorry for the Tommy infantry away back. Within five hours, the majority of them were gasping with thirst and of course their division never got within shot of the enemy. The Anzac Mounted Division had fought the last two Romani days on one quart-bottle of water per man. But we have long been desert wise, splendidly trained, as hardy and cunning as any Arab. The poor infantry chaps, especially the 42nd Division, are in very different plight. Lots of them appear to be just fresh English lads.

The 52nd Division are proved fighters, they have been stationed too at Romani some time. They made a much better attempt, but their orders came too late. There is no doubt that trained horsemen are far ahead of infantry in open warfare.

In the afternoon we watered the horses at a handy oasis. Two troops of us were clustered around the well when there came a screaming whine— bang! directly above us with the hiss of shrapnel-bullets. Buckets were dropped, men scrambled up from the wells, bits were hastily pushed in the horses' mouths at that screaming whine and bang! and we were away in a galloping scatter. Bang! bang! bang! bang! The shells shrieked overhead, bullets hissed between Stan and me biting the dust beneath our horses' bellies. I hunched my back waiting for the bullet to strike and was scared stiff. But we got safely back to the regiment.

Then the Tommy battery that has adopted us, answered "Bang! bang!" and away screamed our respects to the Turkish guns. They replied in kind and we listened to the scream coming, coming, coming—bang! swish-sshssh-ssh—and the rain of bullets churned the sand just behind us and the battery. Then again comes that nerve stretching scream, coming, coming here—ah! too far to the right thank God! as a cloud of smoke drifts lazily over that little ridge. Look out! here she comes again! Men crouch low, the shell screams overhead, bursting in a rain of bullets behind and in among the last of the battery horses.

What an expert range the Turkish battery has got! If they only burst their shells one hundred yards shorter they will sweep the whole of the regiment and the battery too.

With a snappy bang! bang! bang! bang! our own battery-shells screech towards the Turks. Our guns are right amongst us; their discharge blasts the very air. The sweat is pouring off the Tommies. As they swing open the breech, fumes swish back amongst them and curl up like steam. The Tommies are great artillerymen. The battery observer is a short distance away, on a little rise, his glasses trained towards the unseen Turkish guns. The observer signals where our shells are dropping to the Tommy major who stands just behind his guns. The observer signals—the major calls quietly: "No. 1 gun—fire !" Bang! whir-rrrrrr. "No. 2 gun—fire!" Bang! whir-rrrrr——

Back swing the breeches with an oily click, a shell is shoved in, the heavy breech swings closed as easily as closing a watch, the gun-crews step back, the gun layer alert for any changing of the sights, the observer signals, the major orders: "By Battery—Fire !" Bang, bang, bang, bang! Whir-rrrr whir-rrrr whir-rrrr whir-rrrr—and the four shells screech away. The sun blazes down and ah! here comes the return—"Lookout!"—Whirr-schsshssh bang! and the shrapnel bursts a hundred yards behind the battery. The Turkish gunners just can't get the happy medium. How we pray they won't!

From away in front and out to the right comes the rattle of rifle-fire, the clutter of machine-guns, the boom of 4.2's. We see in the distance columns of sand and smoke spurting into the air where the Turks are using high explosive. But we are not in the fight today.

Some man starts to laugh, and pulls himself up sharp. It did sound a little bit hysterical; but it appears that he read an Australian newspaper that said the Turks were short of ammunition!

With what extraordinary fortitude our horses stood that infernal racket On the Peninsula, I used to imagine that horses would go mad when under shrapnel-fire, but here they were with it bursting right amongst them, our own guns blazing away under their very noses, and only by the lifting of a few startled heads, just a little uneasiness as shells screamed exceptionally low overhead, did they betray perturbation at all. Then they would settle quietly down again.

Bang—crash! Christ! showers of sand, plunging horses, a gaping hole within six yards of C Troop—the shell had skimmed down right over the horses' backs! Had it burst one half-second sooner it would have wiped out the whole of the troop. How easy for us clustered mates to have been lying just there! I picked myself up, trembling as I wiped the sand from my eyes.

And so, for the rest of the afternoon the Turkish shells kept bursting only a few yards too short or a few yards too far behind the regiment and all around us in a circle, and missed the tiny hollow in which were crouched six hundred men, and horses and the guns. We only had two wounded and a few horses killed. I know now that miracles happen.

A staff officer came riding up and told us of a mobile column under Colonel Smith, V.C., who apparently are indulging in a little private war of their own away out on the right somewhere. It is the first we have heard of them. They are a mixture of Australian Light Horsemen and British Camelmen. They seem a cheeky crowd, out in the desert on their own without any supports, butting into any Turks, big or little, wherever they find them.

The blasted taubes came hovering over us, seeking to locate our exact position so as to drop smoke-bombs on us and correct the Turkish batteries. We felt utterly helpless against those dreaded taubes, for our position then was that we could not move. And yet, in the sky straight ahead, each time that one of our own 'planes made a hurried dive over the Turkish lines there echoed the sharp *crash! crash! crash! crash!* and a battery of anti-aircraft guns sprayed her with shrapnel-clouds. But England had no anti-aircraft guns to protect us.

Thankfully we saw the sun set. The troops were to retire. Our regiment retired, a squadron at a time, but dark as it was the splendid glasses of the Turkish observers saw us cantering over the skyline and we retired to the crash of their following shells.

It was quite dark when we rode into El Katia. In amongst the big palms we watered the horses, fed them. and wearily slung off our fighting-gear. Brigades and regiments and batteries were rolling in. To our surprise, those already assembled had the whole great oasis a-twinkle with fires. Fires under the very mouths of the Turkish guns! and yet if, in distant Dueidar, we so much as lit a pipe after nine o'clock the Heads would go mad.

It was well for us that the Turks were getting such a thrashing, that they took advantage of every chance to retreat, seized each momentary respite.

So we lit our fires too with delight, and each section put on its quart-pots. There was plenty of wood, such a luxury in the desert. We got real hot tea into us, and with bully-beef and biscuit we lucky ones not on duty rolled back on the sand and fell fast asleep.

Long before sunrise, fires were already twinkling. We lit up hurriedly, afraid lest we be stopped. A drink of hot tea when a man knows not what is before him, is great. The east was greying as the column rode out. We wondered if we would get "into it" to-day. The feeling was not pleasant.

The regiment and guns moved to about the same position as yesterday. Word was signalled back that the Turks still held Oghratina in force. Rifle-fire was crackling in the early morning, guns were booming. The regiment settled down, awaiting the order to move up into the line. C Troop were sent out some distance as a right flank guard. We spread out in artillery formation so that the shells would not get us all in a bunch. Stan and I rode together, Bert and Morry about a hundred yards behind…

We knew when the Turkish observers sighted us: Bang! whir-rrrrrr. We were only a score of men but they lathered the shells into us. We had very little cover. Some of the shells burst right in front and I noticed now that when there were only a few of us and well scattered, the horses were much more frightened; they shivered at the whine of each oncoming shell and plunged violently should it burst directly in front. Some shells threw a cloud of black smoke with a rotten smell, and these particularly frightened the horses.

Suddenly I noticed where some C Troop men were standing, right in front, a spurt of sand. Then another. A sniper! We scattered as if from the devil—only we dare not go far without orders. Rifle-fire was rattling just in front, but I could not hear the sniper's rifle. The lurking terror must have been using a silencer. Then plop! a bullet landed between the heels of Stan's horse. We hurried the horses behind another sand-mound. It was nerve-racking, being fired on by an unseen and unheard foe. This kind of fighting is far worse than trench warfare where the sheltering trench allows a man to shrug at shrapnel and bullet.

Later, we were ordered to line a small ridge away to the right. Gladly we galloped across, for here was cover of sorts. Stan and I dismounted beside a conical mound and lay down by the horses, watching the shells bursting all around. The Turkish army had done a wonderful thing crossing a hundred miles of Sinai desert dragging guns and stores. How they had brought along the vast quantities of ammunition and heavy guns, had even our Heads completely puzzled.

We soon spotted a cunning Turkish trench only forty yards away to our right. It was shallow, in a commanding position encircling the crest of a ridge, each man's possy so naturally hidden by the bushes that it was impossible to distinguish it until ridden right up to. I shivered. If it had been manned, there would be no Troop now. The Turks have dug hundreds of those trenches to cover their retreat. We lay gazing at that trench for hours, expecting the rifles of snipers to come peeping through the bushes.

Stan pointed away out to the right, where shrapnel clouds were drifting over the 3rd Brigade. The Turkish army was beaten, but it still had a sting in its tail. Presently we watched some big New Zealand patrols pushing among the sand-dunes, chased from hill to hill by the shrapnel, the shells searching unseen gullies where men might be sheltering. We were expecting every moment the order to get into action. The sand was scorching hot on the white sand hills where no shrubs grew, the glare seemed reflected from a steel-blue sky.

To our delight Stan found a tin of milk in his haversack. We opened it with the bayonet and had a great feed of bully-beef and biscuits. The horses hungrily tried to eat the biscuits from our hands. My little pony bent his head and gently nuzzled his lips against mine, his big brown eyes saying plainly: "Give us some crumbs."

In the afternoon, we thankfully left the new Relief to the shells and returned to the regiment, which was now catching it so heavily. But just before sundown there came a tornado of shells smartly answered by our little spitfire of a battery. Then came high explosives striking the ground with a whining crash chorused by the humming fragments of shell. Soon the battery was enveloped in clouds of sand and thick, evil-smelling smoke. We just hung on to our horses and waited, for there was nothing else to do. How we watched the setting sun!

When the order came, we were in our saddles instantly. The order was to retire, one squadron at a time, five minutes' interval between each. A Squadron was the last to retire. The waiting minutes were very long. C Squadron trotted smartly out, the Turkish shrapnel following them over the crest of the hill. When safely away, B Squadron followed, riding rapidly to the right, to the roar of our own guns and the whine of the Turkish shells. But the few seconds necessary for the Turkish observers to see the change of direction and alter their range accordingly was sufficient for the galloping squadron to get away.

Our turn came. With intense relief we moved off, only to halt just over the first rise so as to confuse the Turkish gunners who would train their guns expecting us to appear on the rise ahead. I glanced behind and saw Major Johnson coolly riding up behind. A flame burst high above him and—bang! a shell-fragment churned the sand at his horse's heels. I laughed foolishly at the old major's quizzical glance around. Then we were off at the gallop and the Turks fired as fast as they could load but the shells burst just too far to the right, or a little too far behind. And presently we rode into El Katia and camp.

23

August 8th—Hod-el-Amara. A fiery sunrise saw the regiment out into the desert; we felt certain of a fight to-day; we knew that the Oghratina redoubts had been feverishly strengthened. As the dim hills took shape that queer feeling crept over me; I believe every man experienced it when anticipating a bullet, expecting the shrapnel. It did not come. We passed our position of the day before and rode over the skyline. Not a sound, not even a rifle shot! We passed lots of snipers' possies, a man found himself imperceptibly edging away from them although he saw they were empty—those cunning holes dug deep down in the bushes. Surely, we thought, the Turks must be waiting for the regiment to ride over some clear ground on which their gun-sights are concentrated; we will get hell presently. But—not a shot. Not even a sniper. We wondered and waited, riding on.

A mile farther, and we realized that they had retired during the night! I was glad, anyway. We rode over their outlying trenches, empty except for cartridge-clips, as silent as the Turks that manned them here and there. We neared those low, forbidding hills of Oghratina. We waited again while riding on. The brigades coming behind, those advancing on the flanks, loomed up plainly—Still not a shot!

We rode over the largest redoubt thankful that we had not had to charge it. The regiment carried on toward Dababis, took a few infantrymen with some German ambulance men, and huge stores of abandoned ammunition and provisions. Meanwhile eight of us under Sergeant Major Hanson were detailed to get in touch with the Turks. We hurried out past the screen, our horses snorting sideways as they trod on half-buried Turks, then heading a little south of east we rode out into the lonesome desert. Gazing back I saw some of the screen gallop away, leap off their saddles, and root a nest of snipers from the bushes.

We eight rode well apart lest snipers make us a group target. We rode in among hills of rolling sand, bare of bushes, which lessened our fear of snipers.

Ours was a risky job. The Turkish army was retreating pell-mell, but we expected the bullets of their rearguard every moment. If so, then any of us who were left would gallop back and tell the regiment that we had "found" the Turks. A man sleeps long who goes to sleep on that job.

About a mile ahead, a vulture squatted on a peak. It stood up—a Turkish infantryman. He watched us, then leisurely climbed over the sandhill to tell his mates. So one Turkish outpost at least knew we were coming!

We scouted cautiously ahead and among the hills came on one solitary date-palm, Bedouin clothes scattered about, earthenware jars among melon-vines, all watched over by a lively young camel, his rough wooden pack already upon him. It was hard lines on the camel, poor inoffensive animal happy in his sunlit desert—I felt like a criminal when we shot him. You see, the Bedouin was hiding among the sandhills waiting to sneak back, mount the camel and hurry off to arrange a Turkish ambush.

We pushed on quickly and read on a page of the desert, the tell-tale tracks of thousands. Tracks, tracks, tracks—tracks of men, of camels, of horses, boot tracks, barefoot tracks, slipper tracks, sandal tracks, tracks everywhere, and a broad road of tracks winding plainly among the limitless hills—the tracks of a beaten army.

Here the Judas desert showed us too just how the Turks had got their heavy guns across those leagues of sand. They had laid planks before the gun-wheels. As the guns passed over, men kept picking up the planks and laying them before the wheels again. What ceaseless back-breaking work! But we knew how effective.

Eager now as schoolboys—the silence seeming to breathe of great events, of life and adventure and despair all around us—we rode expectantly on, nine men in the tracks of an army. Presently we came on a deserted camp. The officers had sheltered from the pitiless sun under huts built of palm branches. An Australian aboriginal would have fashioned a better gunyah—but then he had the material! The sun glistened on empty bottles, fish and meat tins.

The moody desert then changed into abrupt hills walled with gold-tinted precipices. Almost every hill-top had its rude outpost shelter, but of living man no sign. Just the desert, and the sky, and the sun, and the tracks of an army in the sand.

Up to our saddle-girths we floundered down a precipitous hill of floury sand to unexpectedly stare down on to a roof of palms far below. The sand will inevitably overwhelm that oasis. The sergeant-major and I skirted the palms, peering into the shadows. Three chaps jumped off their horses and with their rifles at the "ready" peered among the gloomy trees, when—crack! Corporal Logie slung up his rifle—up flashed the other two, and "crack crack! crack!" sang through the palms. The sarn-major and I galloped around the oasis to prevent anyone escaping—there might have been a regiment hiding there for all we knew!

In the heart of the oasis we saw them—we galloped in yelling "Stanna,

Jacko Stanna! Stanna!" One chap was kneeling behind a palm frantically waving a tunic, screaming something that sounded like "Yah! Yah! Yah!" The other was down, his face all screwed up—a great fear in his eyes. I thought he was only frightened, and laughed at the strange Turkish sounds he was making. Afterwards I was very sorry; the poor fellow was shot through the arm and clean through the body, fatally. We bandaged him up as best we could and left his water-bottles filled close handy. The other man would have told us all he knew had we been able to understand. They had been eating green dates. They were snipers, their ammunition belts were half-empty too. One had a little aluminium flask half-full of brandy which we left beside the wounded man, but just as we were going away I'm damned if the prisoner didn't stoop down and pocket the flask. We sent him back with "Uncle" Meiklejohn, who if he could find the regiment would tell the doctor where the wounded Turk was. As his mate was marched away the wounded man, who had been wandering in his head, tried to sit up and cried out in Turkish. We were very sorry, but had done all we possibly could. It was the fortune of war.

We breathed freer when we climbed to the circle top of grim hills enclosing that nameless oasis. To have been caught down there would have meant a death-trap. We rode on, alert for fight or flight, but we only saw tracks, tracks, tracks.

We munched a biscuit going along. I felt glad I wasn't the Turk away back there in that lonely oasis with a bullet through his lungs. I turned stone-cold when I thought of Bedouins—the Bedouins torture friend and foe.

Presently, the desert became more open, hills smaller; bushes appeared, then small hills covered with bushes, and our dread of the sniper returned.

Some distance away we espied a large oasis and three of our fellows rode into it on their own. We were getting jolly anxious when they emerged with a young Bedouin girl. We were amused and interested. Apparently about seventeen, brown as a berry, she was pretty in a frightened, sun-scorched sort of way. She was wild as a desert rat. Her hair was jet-black and extraordinarily long and I'll bet it would have taken a horse-rake to have combed it. She tried hard to tell us which way the Turks had "Emsheed!" Her flashing black eyes were nearly out of her head as she tried desperately to betray the men who had been her friends but a few hours before. She appealed to me in a volume of gibberish and pointing of little brown but dirty fingers. I grinned and said: "Quack! Quack!"

"Damn you Idriess!" said the sarn-major.

Bob Hanley started off back to the regiment with our "desert queen." He waved his soiled old hat to our congratulations.

We carried on again, quite cheeky in that we had the desert all to ourselves. We knew that there were thousands of men marching somewhere behind us, and other thousands just on in front, but what happened to those parties of course was their concern. Presently we climbed a commanding hill from where we could gaze across the desert for miles ahead at that great track vanishing into the hills. But of live Turk, not a sign.

The sergeant-major then turned nor'-west so as to cut the regiment, for we had conclusively proved that the Turks were retreating rapidly and were well ahead.

An hour later and we scattered at sight of a hostile looking crowd. Both patrols advanced in skirmishing order, but the other turned out to be a patrol under Lieutenant Wood. They were hungrily chewing luscious dates and what appeared to be greasy rolls of brown paper. A captured German officer and a few Turks stood glumly by.

We hurried towards the abandoned camp, watered the horses and rushed the stacks and stacks of food-stuffs there. Hundreds of bags of grain and flour and cases of tea and goodness knows what else. What we rushed most though were the sacks of dates and huge rolls of apricots, all compressed like layers of thick brown paper.

What a gorgeous meal! It was a feed for the gods, after being so long on bully-beef and biscuits.

I packed up my canvas water-bucket with a surprising quantity of dates for Bert and Stan and Morry, and tore off a roll of apricots. The loot was heavy for the poor old neddy but the section would eat me if I brought no spoil back.

We went searching for the regiment and crossed tracks where the Turks had dragged their guns by packing the wheel-ruts with bramble and bushes and sand all rammed tightly until both wheel-tracks were solid and hard.

Farther south, we came on the troops, brown columns of men winding among the gold and brown of sandhills. The hills away ahead were spider-webbed with the screens, followed by the advance guards of the brigades. The main body marched ceaselessly by to the accompaniment of one mighty murmuring, that was the horses' hooves in the sand. Clearly came a snatch of laughter, the call of a man to his mate, or an irrepressible Australian whistle. As a column of big New Zealanders rode by, the sergeant-major told us to wait until he found the regiment. The En Zeds were positive that the regiment had halted outside Oghratina. We shook our heads, and waited. Men and horses, guns and wagons kept rolling by. Time passed, the sarn-major did not return, we thought he must have missed us among all those troops. Regiments passed who knew nothing of the 5th.

Finally we determined to go back. What a weary ride! Mile after mile, weary horses, weary men, and the troops passing us going in the opposite direction. At last we reached Oghratina and of course found that the regiment was far away ahead.

The salty water at Oghratina would have driven the horses mad. We had to push right back to El Katia—the setting sun made the desert inexpressibly gloomy. By Jove, the horses did drink. Even when full they gasped and stood with open mouths in the water-buckets though they could not drink one more drop.

We swore good-humouredly, gave them bulging nosebags of grain we had pinched from the Turks, then wearily boiled our quart-pots. After a gorgeous tea of bully-beef and biscuit, date and apricot, we sprawled on the sand to gaze up in surprise when a troop of English Yeomanry rode up and asked if the water was safe to drink. They seemed utterly weary.

Their two officers, typical Englishmen, strolled over and had a yarn. They were very decent chaps. We soon noticed that the Tommies were eating nothing. They sat on the sand or lay around in dejected groups.

They had been riding over the desert since the night before. Of course we ransacked our haversacks and gave them all the bully and biscuits we had, and nearly all our dates and apricots. I just managed to save a handful each for Morry and Bert and Stan. Between the lot of us we managed to rake up quite a respectable meal. Of course, the Tommies were jolly grateful; they bucked up wonderfully as soon as they had a good meal inside them.

After an enjoyable smoke-o we said "So-so" to the Tommies and rode back to Oghratina where we camped until two o'clock in the morning. Then in the dark we rode over the road which had been churned by two armies only a few hours before.

24

Two hours later, we plunged down the slope of a razor-back, staring at a long shadowy smudge in the valley below. As we listened, there came floating up the noises of many horses and wagons. We floundered down with our horses' bellies scraping the sand and gladly found the smudge to be our ammunition-supply column moving up in expectation of a big fight on the morrow. This news hurried us. Soon we left the slow-moving column behind.

August 19th Hod-el-Amara—The east wavered steel colour, softly clouding with pink. We were wondering where the hell the regiment was when Jack Meiklejohn's horse knocked up. We halted, and immediately heard the hum of voices away on the other side of a black razor-back. We started clambering up. "Halt!" sharp and decisive in the stilly dawn. The New Zealand outpost, muffled giants in their greatcoats, were suspicious and seemed rather ready with the bayonet, but we urgently made ourselves known and they pointed down in the gloom to where the regiment was just moving out. The En Zeds grimly told us that the artillery road we had been following led directly to Bir-el-Abd and the Turks only three miles farther on!

To our delight, the regiment actually had BREAD, and a little jam and plenty of bully-beef and water. We made a hurried breakfast, mounted, and quickly caught up our individual troops. Riding along munching the apricots and dates, Morry explained between bites that the Anzac Division was to smack Jacko. The New Zealand Brigade that we were still attached to was to thump him on the nose, the 3rd Brigade was to plug him under the left ear while the 1st and 2nd Brigades bunged him under the right. The 5th Mounted Yeomanry Brigade was in reserve to come up and kick him in the ribs when we got him down.

Which information Morry enjoyed with the dates. He was still sucking stones when the first rifles cracked.

The sun sped up and showed us a panorama of low sandhills, grey-green with shrubs. Soon the clear morning air was ring-ring-ringing; soon came the old familiar *zip, zip, zip, zip, zip, zip,* and overhead the *zzzzzzzip-zzz-zip-zip, zzzzz* of long-range machine-guns.

And my belly responded with that nasty sinking feeling. I've always had a dread of being hit in the stomach or the knee.

The Auckland and the Canterbury Mounted Rifles got straight into bolts with the Turk. The En Zeds are first class fighting men and I don't think they grumble as much as we do. They shave oftener, anyway.

The regiment extended and dashed off over a skyline then down a sloping hill. With the movement of the horses that scary feeling is not nearly so bad. The air hummed viciously, but we moved too smartly for Jacko, trained fighter though he is. He was dealing with expert troops. We were a quarter-mile nearer him and under cover while his gunners still sweated to get our ever varying range. We abuse our Heads a lot, which is quite as it should be, but when it comes to heady points of fighting they certainly know their job.

Then came another gallop forward. Again the Turk missed the range. Again we were under cover, this time much closer to him.

Lieutenant Cain's machine-gun section with Lieutenant Boyd's troop galloped into action beside the Auckland Rifles. The machine-guns were almost immediately stuttering raucously. Major Cameron galloped C Squadron to the flank of the Rifles. The remainder of the regiment waited.

The firing swelled to a growling roar, breaking out in fresh places among the hills and rippling along as regiment after regiment galloped into action. Wherever we looked on any skyline the ridges were tipped by bullets.

Staff officers were busy on the ridge before us. Telephone-line layers were rushing about under fire, keeping that thread of copper wire, of such vital importance, in constant repair. The phone men had the day of their lives that day. I wondered that they found time to breathe.

Then that damned, cheeky little battery of ours said: "Bang bang!" very snappily. In resignation we awaited the reply.

"Boom-oom!" distantly, then a rapidly advancing, hissing scream and—bang! a cloud of smoke, rain of pellets and Jacko had replied.

He poured it into us, our own battery answered faster and faster, the shells screaming over to the screech of Jacko's answer. It was astonishing how our horses stood the infernal din, not a whimper even though now some of them were being hit.

Then the cursed snipers sneaked around and got at us. We'd hear a heavy smack! and know a horse had been hit. The poor brutes mostly got it through the stomach. Some of them, apart from the shivering grunt, hardly moved, others shook themselves a little—it depended on where, they were hit. One reared wildly and pawed the air. Another plunged yards forward on its knees, blood pouring from its nostrils as its head lay over on the sand but mostly they were hit through the stomach and would just shake themselves a little. The owner would take the saddle off immediately, for it was always a case. The horse would nose around among his mates, shake himself, and five minutes later roll on the sand. It was the beginning of the end.

How we hated those snipers!

This time we galloped off from under the roaring muzzles of our own guns. The horses got excited, we got excited, we leant over their necks and shouted and laughed in a mad exhilaration. The Turks could see us and they hurled over the shrapnel until the very air seemed one long scream all rent with bursting shells. The bullets hummed faster, our own guns roared faster, the horses galloped faster, and out in front of the regiment, his tail mast-high, head erect, neighing frantically, there raced wildly excited a long, bony horse which had been hard hit and should have been dead an hour ago. He was having the last grand gallop of his life.

So we thundered to cover again but the wounded horse raced straight up on to the skyline where he pranced in a circle, neighing the regiment to come on until abruptly he dropped.

We dismounted and waited in that awful inaction again, watching the telephone men galloping up their precious copper wire under fire. They rushed it to the staff officers who again were very busy almost on the skyline, almost up to the dead horse. Jacko searched the back of the hill with his shrapnel and we pressed close against the hill endeavouring to dodge the hissing pellets, while now the battle roared directly out in front of us. There came a splitting succession of crashes just above and there, swooping very low over the Turkish lines was one of our own 'planes, very pretty in the early sunlight as she glided in and out amongst those little fleecy clouds. She circled around again and the crash! crash! crash! crash! was an orchestra all to her pretty self. She circled again and flew higher and came serenely back with bursting shells all round her. She was so low, at such short range it seemed a miracle she was not hit. Then she swooped down towards us, so low over the massed horses that we could see the goggles of the aviators as a hand appeared, then a long ribbon came twirling and falling from the 'plane as the message dropped right in amongst us. We cheered the two men as they waved and rapidly flew higher.

By Jove, the sand got warm. We chewed a piece of biscuit, but saved our precious water. Then "Mou-nt!" and we were galloping over to thunder down the slope away across a flat and rein up our panting horses under a hill very close to the Turks. "Dismount for Action!"

We flung our reins to the horse-holders and lined up one little line of men, the knowledge that we were going towards death making us intensely alive. The colonel said a few words in that quiet old voice of his, then handed A Squadron over to Major Johnstone. The old major looked at us, then said quietly: "A Squadron, Advance!" He turned and walked off towards the Turks, we followed.

A New Zealand machine-gun started blazing away beside us, covering our advance. Her stuttering bursts leaped into one harsh outpouring, we could tell by the sound that her lightning mechanism was thrashing out against time and heat and sand. Easily understood how on that day the barrels of our machine-guns melted under their own fire. Very soon, steam was curling about her sweating crew as the water in her cooling-jacket came to the boil. Throughout all our long advance, her bullets, four hundred and forty every minute, streamed over us unceasingly, aiming for the heads of the Turks who were aiming at us. You see, it takes good men who, their heads being sprayed by a machine-gun, can lie still and air coolly back in return.

"A Squadron, extend in skirmishing order!"

That was a great advance towards those sombre hills we would never have got there so fresh but for the cool, thoughtful leadership. "Lie down and take it easy, boys, let no man hurry," shouted the old major and the line of men sank gratefully down on the blazing sand, getting back their wind in deep breaths, wishing and wishing we had more water. Then on again, every man hurrying to rise at the order, plugging along several hundred yards farther, the major gradually edging the line on a slight, unhurrying turn until each man was marching in a direct line for his distant position. Then down again, another welcome spell. Up, and on again, the weight of the ammunition bending our shoulders, the sand gripping our feet. And the hissing, the spurts of sand from the cursed snipers' bullets coming in from our flanks. Then spell again until gradually we got our second wind and marched easier on towards that dark skyline that seemed so far away. Suddenly "Lie down men!" We gazed to the skyline expecting the order to fire, then wondering, staring. Silhouetted on a sandhill top were two lines of men in violent action running forward firing from the shoulder, clashing, we could see pin-points of steel, men jumping back in short halts, lunging forward, firing, rushing forward again, clubbed rifles, faintly we heard shouts. Presently they sank down again. Were they Turks? And yet they were on the same position the New Zealanders were holding! The old major put down his glasses, looked over his shoulder and called out: "The New Zealanders are engaged in hand-to-hand fighting, boys. They have just repelled a charge. They will be very tired. We must get up to them and join in. Advance!"

Hand-to-hand fighting in such heat as this! We wondered the men's hearts did not burst. We panted on.

The rifle-fire was a harsh roar as we drew nearer, the shells burst with that rending crash they have when very close. We stepped over some dead Turks and a New Zealander—face down in the sand. We climbed and puffed our way just up to the skyline and looked down on the fight. We were joined for a while by the colonel, who had just placed his crowd in the firing-line. Since the fight started at dawn the Turks had been pushed back losing three circles of hills. They were now at bay.

The bullets whistled viciously at close range. We advanced in short rushes across the broad, flat crown of the hill, from cover to cover, a line of men in section groups for thus we could avail ourselves of the irregular sand mounds. A lather of sweat, we came to the rim of the hill and crawling gaspingly forward sank down behind the last circle of mounds on the very lip of the hill. I laid my rifle in the bush before me and gazed straight down into the Turkish position.

It was a broad flat, dense with bushes, swarming with Turks. The flat, almost a plain, was encircled by the hills from which the Turks had been driven. Around Abd were big redoubts and trench-systems that Jacko had been preparing these months past. Nearly straight ahead, not three miles away, was the oasis of Bir-el-Abd, its trees a great dark patch. A little to the left, was the Turkish Red Crescent. Around their main camp by El Abd itself the sand-dunes were bare and shimmering, but the great flat before us was dotted with a million sand mounds thick with bushes. 'Planes circled in a cold blue sky, puff-clouds not of nature drifted high up everywhere.

Spreading over the flat, where the Turkish battalions were thickest, drifted haze-clouds of shrapnel. In the bursts of quick succession right above the Turks we would see them running for cover, we could distinctly see those that fell and lay there. It gave us a strange feeling of satisfaction, lying there getting our breaths, gazing at our own shrapnel bursting among the Turks—they had given us such a lot of their own.

I thrust out my rifle. The barrel was already hot and I had not fired a shot. But now there were plenty of targets.

25

It was good shooting. The Turks were advancing again; as they left shelter on their short rushes forward we gave them what they had given us—and their covering troops poured the lead into us to keep our heads down. Bullets whipped the sand—clipped twigs from the bushes—a spray fell across the rifle-sights just as I was about to pull the trigger on a man, and I cursed. Their machine-gun bullets clipped the bushes, ripping them to their very roots. Their German gunners and Austrian artillery-men are horribly efficient. In miles of a semi-circle crowning those hills we lay, clustered in sections behind the sand mounds, because the clear sand spaces in between were dancing from the machine-gun bullets. Soon my ears were ringing and ringing and ringing from the concussion of my mates' rifles, for we had to fire with our rifles close to one another's heads. We fired until the rifle barrels grew so hot they burnt our hands. I rinsed my throat with hot water—no man had a cool drink that day.

The Turks came on with an inspiring bravery, it was their second counter-attack. They suffered dreadfully, they just seemed to dwindle away. Then they began to retire. We were ordered to link up closely with the New Zealanders. We did it rapidly, but it was awful, gasping under the heat, the short rushes, the fury of bullets immediately concentrated upon the bare crown of the hill. Simply a rain of bullets and shrapnel followed us. A man's feet felt clogged with lead, the sand rose up and down in simmering waves, I fell behind a bush and felt I could never, never move again. But quickly we were blazing away down on to the plain where men were running back, farther and farther, and now in little groups growing larger until their whole line broke and away they went for El Abd, all broken up, running bent low to the earth. And we gave them hell.

At last they got back to extremely long range. They stayed in the bushes and the shrapnel stayed with them in hazy wreaths of smoke. And in the interval, breathed and looked about us.

Here and there lay our poor chaps, New Zealand and Aussie, but many dead Turks lay among them, and a lot of our poor horses, too. The New Zealanders had it the worst so far. In the first counter-attack, two columns of Turks easily six thousand strong had advanced against them alone.

We had hardly got our breath when again the bullets rained like a stormy shower. As our men were wounded a mate would pick him up and stagger back down the hill There was no cover for those who carried that worst of all burdens, a badly wounded man; they just walked laboriously in the awful heat over the sand patches whipped by bullets. We got all the wounded down to safety like that, lots more were hit doing it. The doctor would patch them up, creaking sand-cart or grunting cacolet would then take them away back, a long ride of agony.

The bullets freshened in showers that told us the Turks had been heavily reinforced. Thousands were pouring in from their base at Bir-el-Arish, miles back along the desert. Our Camel Corps had quite evidently failed to stop the new battalions.

The Jackos were actually running towards us now along a central front of three miles and as they advanced their comrades fired over their heads a covering fire which crackled and strengthened until the hot air was hissing, bushes were cut and slashed and twisted, the plop, plop, plop, plop of bullets landing amongst us united with the shrapnel-bursts in racking noise.

We fired until the rifles were again too hot to hold, until our elbows were blistered in the sand, until the cry rang out: "More ammunition!" "Fix—bayonets!" called the old major, "Prepare to repel the Turkish charge!"

A queer feeling ran through every man of us as each lay over on his side and wrenched the bayonet from the scabbard. A totally new sound, a quiet little sound whispered all along the line, "click, click, click, click," as the bayonet springs clipped fast.

I stared to right and left and there among the bushes in line was the flash of the grim bare steel and a drawn, angry look on the faces of the men. A strange thrill ran through me—nothing in peace-time could make a man feel like that.

Their survivors came right to the very foot of the hill, they hesitated, they stopped. I was not feeling brave myself, but I smiled. A man only had to think! Had the Turks charged up in that frightful heat, those that lived would have collapsed at the top; I hardly think a man would have stood on his feet to meet our steel.

But their battalions melted down there and directed a furious fire at us shooting from the crest. And just then panting men came staggering along our line and as they passed they opened haversacks and threw down ammunition. We wolfed those cardboard packages, tore them open, filled our bandoliers and belts and blazed away again. One of the ammunition carriers crouched exhausted beside our mound, volunteering some hoarse gossip. "Lieutenant Wood and Graham's just been knocked. Th' 3rd Brigade is held up by th' big Barda redoubt—th' Heads have got from the prisoners that th' Turks outnumber us three to one, apart from th' reinforcements hurrying in from El Arish—th' taubes are chasing our 'planes away … You oughter see th' Tommy batteries, th' Somersets an' Leister an' Ayrshire—they're workin' their guns in their shirt-sleeves with their chests all red-raw from the sun. A 4.2 landed on th' Ayrshire wagon-team, blew four men an' thirty-seven horses to hell. Th' Warwickshire Yeomanry have been sent in on the Noo Zealanders' left; they've just repulsed a 'Allah-um-Allah' charge. Gripes, I'm thirsty!"

Then the Turkish batteries, to keep our heads down while their infantry continued the charge, rained shells upon us until the hill-top was crowned by a moving, fleecy cloud. Then came coughing down the two first high explosives to burst in an ear-splitting crash! crash! that spouted the sand with dense, evil-smelling smoke. So they got the range and then crash! crash! crash! crash! crash! crash! crash! crash! firing in batteries until we could only see yards away through the drifting, rolling clouds. Our hill-top was churned into a volcano, screeching with flying fragments of shell. For every shell our guns fired, the Turks fired twelve, and much heavier ones. Even in that inferno, we could tell that the New Zealanders and our own C Squadron were getting the worst hell. Even among the hidden slopes the Turks sent their searching shells until the ground shuddered, while again and again rolling black smoke covered the Turks from our rifle-sights.

Presently I was in such a state I didn't care a damn. Glancing over my shoulder at the increasing shell-holes I realized how miraculously close they could burst to a man without killing him—the loose sand saved us, for it smothered the flying splinters. Had the shells been falling on rock—Anyway I knew if a man was to be killed then he would be killed, and that was the end of it. Concentrated shell-fire when a man is lying out in the open is just frightful. And yet the men fired through it all with a deadly, cool steadiness: that unflinching shooting has saved us again and again. Occasionally I glanced along the line at what faces I could glimpse. Each face was terribly concentrated on the job just to shoot straight and shoot again.

Rifle-bolts clicked distinctly, jerking out the empty shell, ramming home the new—as definitely and remorselessly as fate.

Again came the cry: "Ammunition! Ammunition!" "Run down for the Camel ammunition!" "Quick, boys, quick!" Happily the smoke eddied upwards from the lip of the hill as the Turks came again. More fresh men, they jumped up from the very centre of the flat, three long lines of them: they looked rather superb. And the covering fire of their comrades swelled to a shattering roar. By Jove, the Turks must have abundant machine-guns.

We had taken off our bayonets, for you cannot aim certainly at a man with a long, heavy knife on the end of your rifle. It was grand shooting, those waves of men who came on and on and on. As their ranks thinned other men sprang up from the bushes and refilled the lines. And I suddenly remarked that the bushes right to the foot of our hills also sheltered Turks! But I only saw one complete wave of Turks reach the base of our hill!

The 2nd Light Horse on our left were getting pure hell, their hill was another volcano crest under the howitzer fire. The Turks, in overwhelming numbers constantly reinforced by El Arish battalions, came straight at them. Stubbornly the brigade fought but we could tell by the intense roar of the fight that they were slowly being forced back. The Turks found that a donga left a gap in our long thin line. They rushed battalions down into it—they began to pour through. They rained lead on to the Ayrshire Battery and down went men and horses in an awful medley of rearing animals, agonized neighs, and cries of men from out the smoke and dust. I thought the guns were lost but the Light Horse galloped up some troopers' horses. We got the order to retire. We would not believe it. Everyone was looking around for Major Johnstone— Lieutenant Patrick under fire was searching the hill for him. We did not know until next day that he had been wounded and carried away by the New Zealanders. A Squadron swore by Major Johnstone after that day.

The persistent order was "Retire! Retire! Retire!" Very stubbornly we gave way—every man had worked himself such a comfy possy deep among the bushes, every man was curiously angry, unwilling to go. We retired in groups, kneeling to fire at the oncoming Turks while other groups ran past to kneel and fire in turn. Grudgingly we fought back right to the led horses. I had a spare water bottle strapped to the saddle. I rushed it. Impossible to describe the strength that water gave me. And there were a few precious dates in the saddle-bag.

Then the colonel came along. He looked furious. We all were; we at least imagined we had the Turks well beaten; we did not know then that Royston's brigades had been pushed back on the right and the Turks were surrounding us. The colonel took us back over the next hill and plonked us down again to wait for the Turks. Old Bert was completely puffed, so he took over the section horses while Morry came along with Stan and me. But we had only gone a short distance when they were both wanted to carry away a wounded man, so with the others I climbed the weary hill into cover, not caring much for the hissing bullets, being too intent on cramming those dates down my neck.

We watched Major Cameron's squadron get nearly caught. The Turks were yelling right on them and he was trying to get his wounded away. The last troop of men leaped to the saddle with the Turks wildly firing as they ran trying to beat them to their horses.

The Turkish observers located the led horses of the 8th Light Horse. A frightful tornado of shells burst upon them, draping the desert bushes with entrails and legs and heads of horsemeat. Horses scream frightfully when in agony. No matter what hail of death is around a man he sort of forgets his own peril, but an unexplainable fear shivers through him when horses scream like that.

The 9th Light Horse were fighting desperately, a squadron lost three of their wounded; they tried to get them away but the Turks howled right on them, one Turk actually fastened his claws in the mane of a trooper's horse and swung into the air as it reared. He didn't let go either until the trooper crashed his rifle-butt into his skull as his mouth was open yelling "Allah!" The 10th rushed up and charged the Turks. They got one of the wounded men; he was desperately hit. The Turks had torn his clothing off and left him in the blazing sun.

But new enemy battalions kept rolling up and we watched the men on our right stubbornly retiring, fighting and retiring and fighting again. So we lay, and refused to scoop a possy in the sand, angry for the holes we had left behind. We waited for the Turks to come and push us off, but they just fired at us from cover.

It was getting late. We had thousands of horses to water—they must drink. For horses in the desert win or lose battles.

Then half a squadron of Warwick Yeomanry appeared and panted down beside us. I was too dashed tired to enquire how on earth they got there. I just breathed on, amused by the three lads who snuggled close by. They argued volubly as to where the Turks were until one man wiped his nose and appealed to me. I just pointed.

They each scratched a hole, one chap swearing in their comical English way when a comrade scooped a shower of sand in his ear. They settled down perspiring side by side. The lad next me was excited, he was only a boy, could hardly have been under fire before. He plugged his chin into the sand every time a machine-gun fired a salvo. Then he slung up his rifle with the muzzle only inches from his mate's ear, and fired. His mate sprang convulsively out in front, fairly deafened by the concussion. He leaped back, mad with rage. The furious argument that followed made my sides ache from laughing. The excitable one was quietened under the order that to fire again meant instant death for him. Then one lad swore he saw a sniper in a bush fully eight hundred yards away. There may have been a sniper there, but if so he was soon dead, for the Yeomanry man fairly pumped bullets into that bush—he proved an excellent shot.

The Yeomanry were very willing; three of the poor lads were hit before they had been with us ten minutes. They did not seem to possess that instinct of fairly worming themselves into cover that saves so many of us.

The machine-guns got some of our fellows here also. Corporal Logie was shot through the head. As the sun sank, lengthening ribbons of flame spurted from the machine-guns. The Turks concentrated their fire on our ribbons. The Yeomanry gave their helmets to the En Zeds who held them over the gun-muzzles. The helmets checked the tell-tale flames but naturally didn't last long.

The colonel gave the order for retirement and I was damn glad. Perishing with thirst, and hungry, we knew it would be long after dark before we had the horses watered and fed and a chance for something ourselves. We shouted to the Yeomanry to come along, then retired in firing groups all the way down that weary hill to our sheltered horses.

We sprang into the saddles, the colonel waited for the last man, then we extended and at full gallop went away over the hills with the crash of bursting shrapnel in our ears.

When at last we had reached a well, watered the horses and ridden back to camp, we were astounded to be told we could light as many fires as we liked. Turkish reinforcements were still pouring into El Abd. In two hours we were to retire back to Oghratina.

Soon we had the desert twinkling with fires. A miracle had happened, for the camel-train had just lurched into camp with bread, bread, bread, and jam, and even condensed milk, and bully-beef. What a memorable meal that was! I was stretched right out, satisfied, half-closed eyes, contentedly smoking the ache out of my body when the shouted order came: "Reg-ment! Pre-pare-to-M-o-u-n-t!"

What a weary ride that few miles' retirement was, with the dull red fires of the Turks behind us burning their stores preparatory to their fresh retreat. Our own eyes had shown us how dreadfully we had cut them up that day. But we retired, tired horses, tired men, the long crawling column in the dark, the commands of strange voices as other regiments fell in ahead or behind, the ghostly shadows of the Camel Corps padding silently beside us, the weary holding oneself on one's horse. How gamely the horses plugged on, breathing deeply, heads bowed. But occasionally a man sighed as his horse fell, unable to go a step farther. The poor chaps in the cacolets swayed through the night in agony. All the wounded who could, clung to their horses. Two men rode with broken thighs.

The 3rd Brigade stayed behind to fight a rear-guard action if necessary, if not, to keep "in touch" with the Turks. Then the order "No smoking, no talking!" and the rest of the weary ride was like a ghostly army marching through silence.

Only it was not silence. My ears were ringing, singing. In the bushes around, whistling through the air, thudding into the sand were bullets, bullets, bullets! Rifles were ringing, pinging, crackling, a shrapnel screamed overhead and burst towards the rear—all through the long, silent ride. All the boys were hearing the fighting. We were dropping asleep in the saddles. A horse would walk up through the column, evidently lost, its rider's head swaying upon his chest, his slung rifle dragging at his side. A comrade would touch the man, he would sit up jerkily to wait in a daze for his section, hardly hearing the low laugh as the shadowy horsemen glided by. I thought, queerly, of another army—a beaten army that in horror and anguish had trod this same Darb Sultani, the "Oldest Road in the World." And I wondered if the Little Corporal had ridden on ahead of the column, his head bowed upon his breast. And I thought with a smile, nodding in my saddle, of the old colonel riding ahead of the regiment. He was about the same build as Napoleon. We have a lot to put up with from the Heads, still I thought we would rather have the colonel leading us than the Little Corporal. And somehow there marched beside us the cruel hordes of Darius and Cambyses the Persian, yes, Alexander the Great with his Greeks had trod this same old Caravan Route, and Anthony the Roman, and yes, the shades of the old Crusaders were there too riding with the Yeomanry. Crash! Crash! How cold and still the desert looked! Crash! Crash! Crash! Crash! The sandhills' shadows were black as Bedouins' shawls. Then as a man would straighten and try to clear his mind "Crack! Ping-ing-ing"—a sniper's bullet!

At last we reached Oghratina, fell off our horses, pulled them into line, fed them, rolled on the sand and listened to the plop plop plop plop of bullets embedding in the sand. In overpowering sleep I heard the plop-plopping all mixed up with the thought of how very, very glad that I was not on duty that night.

26

The regiment was awakened at four o'clock. I wondered had we really slept! With sleep-aching eyes we fed the ever hungry horses, boiled the quart-pots, ate ravenously, climbed into the saddles and rode straight back after the Turk.

We halted near Dababis, surprised at the order "Off saddle!" We spelled the horses that day, listening to the New Zealanders, the 3rd Light Horse Brigade and the Yeomanry pounding the Turk. Jacko's huge camel-trains were hurrying his wounded and stores away back to Salmana. It must be sheer despair for a retreating army, both fighting and working night and day, mobile horsemen thundering at their heels.

The 11th of August was a happy day—for the section. The regiment rode out, the First rifles snappy and clear in the dawn.

The 2nd Brigade was moving somewhere to our left amongst the sand-hills. The section was sent out to find them, so trotted away, well content at this little stunt on our own. We pulled the horses up into a walk and dodged out for a mile, edging into the desert. Every here and there lay dead horses and in a tiny oasis, a blood-stained hat with the poor owner a few yards away. Dead men lie so still in the desert. In a bush-shaded gully were many dead horses and some New Zealand dead, stripped of their clothes by Bedouin or Turk. We rode on, sharply alert for sniper or Bedouin.

Then, of all things, in an oasis we came on four longhaired goats, tethered to the palms with roughly made Bedouin rope. We were delighted at this fresh meat, but the meat did not reciprocate. Earthenware water-bottles, jars and a medley of Bedouin rubbish was scattered about. Some melon-vines too, but to our sorrow the fruit was not ripe. Fresh tracks were plentiful. The troops hate the Bedouins, ghouls of the battlefields.

A mile farther on, we met two En Zeds riding in with Turkish prisoners. The En Zeds pointed out the whereabouts of their brigade to us. We rode up to the Wellington Regiment and delivered our message and then rode away in high spirits. Presently we met one of our roaming regimental patrols. They told us the regiment was being shelled, but probably would not go into action that day. The patrol scouted off and we took it into our heads to climb one of the big hills of yesterday's fight and have a peep at the Turkish camp. Very cautiously we climbed the hill, past some poor stark Aussies and New Zealanders and numerous Turks, and reached the shell-holed crest of the hill. We crawled forward on hands and knees; the bushes seemed lonely and hostile with our own men not about. We peered away towards El Abd. We could see camel-train after camel-train, packed with stores, leaving the oasis. They were making good their respite while shelling our annoying troops. Rifle-fire was crackling away to our right.

Morry nodded upward, and we gazed at one of our 'planes circling over El Abd. As if dropping from the little clouds above her a Fokker streaked down. We heard the machine-gun fire as the Fokker swooped upward and vanished in the clouds. Our 'plane wobbled dizzily and vol-planed back towards our lines, hit badly. Later we learned that one man was killed and the other seriously wounded.

Presently, as there was nothing doing for us, we sneaked away, mounted and went for a good long cruise entirely on our own, scouting and then galloping into every oasis we came across. We got no prisoners, only old Bedouin belongings hidden under bushes.

We hunted for watermelons and ripe dates. We boiled our quarts in a quiet little oasis and enjoyed our bully-beef and biscuits, but we were hungry. We felt we ought to find something on a day like this. In this particular oasis grew fig-trees but, like everything else in the blooming desert, the figs were poor.

That was one of the good days the four of us had together. We were glad to be alive, glad to be in each other's company.

In the afternoon, we rode back for the goats. Silly old Bert let the damned things break loose, he laughed—I swore and galloped after the goats going hell for leather towards the Turkish camp. I chased one for half a mile, then jumped off and fired—blessed if I didn't miss. Feeling like a goat myself I galloped after the fleeing thing again—it was in full sight of El Abd before I could run the horse over it. Stan came galloping up and slung the panting thing across his saddle. We set off lively, expecting a shot every second, but I suppose the Turks were far too busy to bother about two men and a goat.

After a canter, we caught up with Bert and Morry having the devil's own job driving their goats. Our horses plainly did not like the goats, they tried to tread on them. We got the mutton to camp just before sundown, killed two straight away and when the regiment rode in, C Troop had mutton for supper. It was a glorious feast. Bert found that the goat which had given me so much trouble owned a lot of milk, so I held the scraggy beggar while Stan milked her. He squeezed out half a quart and it went grand in the tea. She kicked a bit.

On the morning that the dawn patrols brought in word that the Turks had evacuated El Abd and retired to Salmana, we were at breakfast when "Taub-e! Taub-e!" and we rushed for the horses too late as through the air came whirr-whirr-whirrrrrr bang! and clouds of sand and palm branches flew around us. Whirr-whirr-whirrrrrr bang! another missed by yards only but showered sand and black smoke all over our bully-beef and biscuits. We stood and watched her other bombs falling among the New Zealanders and headquarters. We could only retaliate with our machine-guns. Had the Taube been one of our own 'planes over a Turkish camp, let alone an army, she would have had modern anti-aircraft artillery blazing into her. But England can't afford us a single gun.

The taube killed a few men, wounded more and killed some horses and mules. A jagged shell splinter pierced my horse's leg, which made me feel just mad. As the regiment was moving out to occupy El Abd, the order came that all men with sick and wounded horses were to march back to Dueidar.

Of course, we concerned started grumbling, but it was explained there would not be much more fighting, the army did not have sufficient camels to carry water and supplies past Bir-el-Abd, and we had completed our job now of driving the Turks out of Sinai. So unwillingly, to the boom of guns, we set back on the long dreary ride to Dueidar, arriving late that night. There were one hundred horses wounded and sick in our regiment alone; it necessitated a troop of men to get them back.

We were all quite anxious to return to the regiment. So before dawn, I took a walk down to where a squadron of New Zealanders was camped. I approached as cautiously as I would have had they been Turkish lines, but a figure rose from the shadows at my very feet and said quietly: "Aussie, clear out of this while you're healthy. We want no more of our horses stolen!"

Despondently I walked back to camp, but grinned on learning the hot reception the other chaps had got on visiting the Scottie officers' horses. A sentry had come straight at them with fixed bayonet and five bullets in the magazine!

We were disgusted and down-hearted. It was very nearly dawn, so hurriedly we crept across to the Indian officers' horses of the Labour Corps. But we found that each Indian groom slept with his horse's halter securely strapped to his arm. We whispered to the horses, and patted them, and started to pull the halters off so we could at least leave something with the grooms. It was delicate work. To our intense disgust we found five of the halters knotted so as to be impossible to undo without cutting. We cut two halters, a man each led a horse gently away. Then a horse snorted, a groom jumped up, they all sprang up yabbering excitedly, but we had gone. In minutes only, the two lucky men had slung on their saddles. As we lay down fast asleep, we heard their hooves pounding out into the desert. The Indian grooms were yabbering frantically, noisier than Turks.

At sun-up we boiled the quart-pots and consulted. Dueidar was far behind the lines, used now as a little base by many regiments where sick horses were brought and where Camel Corps men and officers supervising the railway line and water-pipe line, and details belonging to various regiments, camped on their various duties.

In another two hours we had a bit of fun. Two Indian grooms, evidently knocked up, came riding in leading two splendid horses, belonging to Tommy officers. The grooms evidently did not know the reputation of Dueidar, for they tied their horses up and lay down under a palm-tree only a few yards away.

Barely minutes after, and Young and Spencer had slipped the halters, whipped on their own bridles and saddles and were away with the splendid animals.

A few hours later they rode back, grinning sheepishly. Appears that when six miles out they were passing a group of camel-men and Indians, repairing the line. The Indians waved and called out some cheerful inquiry. Our fellows all unsuspecting and whistling merrily, turned across to them. Then two Tommy officers appeared from behind the camels and one called: "Hello, Prince!" and Prince pricked up his ears and looked pleased while the officer snapped, "That is my horse you are riding!" and the other officer chorused: "That is my horse you are riding! How dare you ride my horse!"

Of course our two fellows tried to bluster it out, but after a heated argument one of the officers said: "Come now, you are caught! Own up." And Young acknowledged: "Yes, we're caught, fair and square."

"Well," said the Tommy, "we are riding those horses back to Hill 70 tonight. You had better ride back to Dueidar with us."

"Right-oh," agreed Young.

The Tommy officers, as soon as they had cooled down, thought the matter a great joke. They all rode back to Dueidar together, Young and Spencer groomed their horses, the Indian grooms indignantly and with commendable promptitude led them away, and we enjoyed a hearty laugh. Eventually I managed to pinch a horse from a man who had stolen it from El Katia. And I've rejoined the old regiment here at Hod-el-Amara, close by Bir-el-Abd. And so the diary is up to date again. It seemed an awful long time writing it up. Just a few of one man's experiences in a scrap. If every man of both armies wrote his experiences of one day's fighting only, it would take a great library to house the books. And yet we hear a rumour that the Government is sending out a journalist to write a "history of the war." What tommy rot!

The fighting, on any scale, appears to be over for the time being. We are just holding the positions until the infantry can come up and dig themselves in.

At El Abd a notice was printed up above two sick Turks "Attention Cholera, with the compliments of the German Ambulance Corps."

A phial was picked up at Geeila Oasis, containing live cholera germs from a Berlin laboratory. We don't know if they were intended to pollute the water or not. Naturally we are uneasy.

The Turks had splendid ambulances. Two field-hospitals we captured were complete with all the instruments, fittings and drugs that science could supply. A captured camel-pack machine-gun company had their equipment thought out to the minutest detail for desert warfare.

At present I'm in the New Zealand field-hospital with something that looks like dysentery. It's been a great chance to finish up the diary.

Some infantry officers were at hospital a while back, discussing the tactical side of Romani with some of our own and En Zed and Yeomanry officers. They were outspoken at the divided commands, Murray far up the Canal at Ismailia, Lawrence twenty-three miles back at Kantara. Chauvel at Romani with all the responsibility but a divided command. When the Turks cut the phone wires, Lawrence of course did not send up in time the troops that were already placed to annihilate the Turks when they had exhausted themselves against Romani. Chauvel asked the 156th Brigade if they would take over the line from the two Light Horse brigades, while the men mounted their horses, swung around and attacked the exhausted Turks in the rear. The infantry brigade was fresh, at greater strength too than the two brigades that had been fighting for a fortnight. The infantry brigadier said that he must take his orders from General Smith alone. Chauvel must have had heartache as he watched the chance of completely annihilating the Turkish army slip by. It is a galling memory to the men though, those thirteen thousand infantry rifles sitting in their huge redoubts, the light swirling right up to them as the thin line of mounted men were driven back and back, and yet not allowed to come out and fight, simply because of a divided higher command and a busted telephone wire. The infantry are as disgusted as we are. One infantry commander on his own initiative sent out several hundred of his men into the fight. The Turks got so close to one of the infantry camps that they plastered their redoubt with shells at close range. No doubt the English have a peculiar way of fighting; thirteen thousand men, fresh, eager for the scrap, and just simply not allowed to go in.

Here in this little hospital too, I hear all about the poor chaps of the 42nd Infantry Division, who a day late for the last decisive day before Romani, were marched the few miles from Bir-el-Nuss to help us in the Katia fight. The East Lancashire lads, quite untrained to desert warfare, were lying exhausted in thousands over miles of desert long before nightfall. Numbers of them were dead, many unconscious, hundreds were mad, digging with bare hands in the sands for water. It is a terrible story, as the hospital men tell it. One battalion alone lost three hundred men; eight hundred men were missing from one brigade. It was only a short march, too.

And now for the sake of possible future memories, and to put in time, I'll memorize. I only saw the fighting, of course, in my own tiny direct front. Only staff officers, possibly, saw it otherwise. In ways it is a peculiar sort of fighting, far worse than trench fighting. Fighting from behind a steel-plated loophole, with your body and head sheltered by a deep trench, is very different to being out in the open and taking all the shrapnel, high explosive, bombs, rifle and machine-gun bullets that are coming to you and the other men. And it is a never to be forgotten thrill a man gets when he actually sees the business-like faces of the men shooting at him. The bullets whizzing by in the open air have too, I fancy, a more frightening sound. The machine-gun bullets certainly have. I know I was frightened many times in the recent fighting.

Against us too, the Turks had more artillery, more machine-guns, more men, more ammunition, not to mention an undoubted superiority in air force. And yet, fighting over the ground with which the Turks have been familiar for centuries, we have completely beaten them and pushed them far back with the loss of more than half their army and vast quantities of stores. Even we troopers recognize clearly that it is only the distance we are now from our Base that prevents us smashing them in their entirety.

The Turks were splendidly led, they fought bravely, desperately, but we are proud of our own men and our own generalship.

Here is what Djemal Pasha said to his soldiers on the 1st July at Beersheba on the eve of the departure of the Turkish Army to the Canal:

> *"Brave soldiers, you are going into the Desert. I ask you to have patience and perseverance. You will return bearing your arms in victory, or you will leave your bones in the Desert. Everything is bad in the Desert, hunger, nakedness, dirt, every privation, therefore, I ask you to have courage and perseverance, O, my soldiers."*

We got the address from a captured Turkish officer lucky enough not to leave his bones in the desert.

But the dearest memory, the memory that will linger until I die, is the comradeship of my mates, these thousands of men who laugh so harshly at their own hardships and sufferings, but whose smile is so tenderly sympathetic to others in pain.

Mounted troops entering the great El Arish oasis.

27

August 24th—Hod-el-Amara. With the old regiment again. To-morrow morning, at two o'clock, a composite squadron of us under Major Cameron are riding out with the New Zealand Mounted Rifle Brigade to reconnoitre north of Salmana to Mat-Eblis and the Island of Galass. We are curious as to the adventure. We have to ford a "lake," and take an island on which is a hostile Bedouin camp. It sounds strange and interesting—"island" and "lake" in the desert!

August 25th—The stunt was a picnic. No sudden death, just a really interesting little trip. Dawn broke in waves of crimson as we neared the coast, riding among low sandhills grey with bushes over which we got an intriguing peep at the blue Mediterranean. Then we exclaimed at a shimmering white roadway fully a mile broad that came in from the sea to disappear inland among the desert hills. Very interested, we pushed on to find that this huge roadway was really an arm of the sea. The column looked quite picturesque riding across that snow-white salt, the horses prancing upon its hardness, the men in great trim, rifles slung, brown arms to the sun, pipes alight, laughing conversation mingling with the snort of horse and stamp of hoof. Then into the hills again and suddenly we were among acres and acres of water-melons. Men jumped from their horses with simply a wild cry, bayonets flashed and water-melons were slaughtered all over the sands. The horses nudged eagerly amongst us for their share, my old neddy pushed his nose in my face and sent me flying while he munched my slice of melon. Bedouin melons of course, miraculously growing in sand and salt. We filled our nosebags and rode on crossing another great arm of the sea, this time covered with a foot of emerald water seemingly resting upon snow. Here and there were what I took to be plump white birds—in reality enormous crystals of salt. The sound of the splashing hooves was exactly like a river rushing over stones. Then we rode on to more melons, scattered fig-trees and date-palms. But the Bedouins had stolen away, leaving only their tracks. We halted and boiled the quart-pots, disdaining bully and biscuits in favour of melons.

An En Zed officer strolled over to us and explained that we were on the famous "Serbonian Bog" of Milton—the Bardawil of the Arabs, called so after King Baldwin the First of Jerusalem, who died here. So our neddies have been treading again right in the steps of the old Crusaders. It appears that by this Bardawil Peninsula, lies Katib-el-Galss, the Mt

Cassius of Herodotus. There was once a mouth of the Nile here, that arm which ran by Pelusium, through which Egypt's trade with the Mediterranean passed, and along which the old galleys used to sail to the Red Sea. The En Zeds had found ruins of the ancient towns Flusiat and Khumiat. Where German officers had recently done a lot of excavating, there were marble pillars with crosses on them, mute evidence that Christians had lived here too. The famous stone of Baldwin was found. It had been carried here, though erected in El Arish by his followers soon after his death in AD1118. The En Zeds had also found that a big hill in front, with the waves rolling against it, had a cliff in which was a layer of Roman pottery, bricks and stone, where Cassius had built a fortified camp. Herodotus had seen the hill in 300 B.C. So that, instead of blowing away, it had added at least fifty feet to its height, measuring from the layer showing the Roman Camp.

It was all intensely interesting, listening there in the desert, our horses munching melons. I hope I have got it down all correct though; the officer was an enthusiast, but I'm very hazy on ancient history.

Some of this was retold us alter the trip, so I think I've got it down all right.

On our ride back we found that the hard salt was deceptively boggy away from the beaten tracks. We returned to camp at 9 p.m. The horses were dreadfully thirsty, having gone without water since the night before, though our squadron had managed to get a taste with the spear pump.

August 26th—The melons have given us an awful bellyache. Our poor horses, too.

August 28th—Our squadron is here at the oasis of Hod-el-Bada. The shady date-palms grow in rows, their bunches of shining red fruit drooping towards the ground. B and C Squadrons are camped parallel with us; we are on the farthest outpost line, to guard the main body against surprise. Everybody is on duty at night, as are the cursed mosquitoes and flies in the day. All these oases were Bedouin living-places. Their goat-shelters are square yards made of palm branches rudely lashed together. Covered by bushes are their earthenware water jars, basins of wood, goatskin water-gourds, camel-skins, dried water-melon seeds, the rudest of rude agricultural instruments, and other rubbish that would have been rubbish in the days when Moses was a boy.

What strange scenes these oases must have witnessed throughout the centuries. They stretch all along this "Oldest Road in the World," starting at Dueidar, going right across the Sinai desert to the Plain of the Philistines, near Gaza in Palestine. The wells are all in the oases; away from them are God only knows what great areas of blazing sandhills. So that all the old Caravans and armies followed the oases along this Darb El Sultani. There is no road of course, camels don't need a road. Kantara is the starting-place, the Arabs call it "The Crossing, the Entrance Gate to the Desert from Egypt." The Pharaohs trod the place, the Babylonians and the Saracens and hosts of others, and now the army of the youngest nation in the world is fighting and riding across it, laughing and singing, joking and swearing and growling, and by now all of us on some night or other must have lain on sand which is partly the dust of some ancient soldier.

Someone was just telling us a while back that the En Zeds have found a long Roman trough built for watering animals, in a beautiful little oasis not far away. The brickwork and plaster are almost as fresh as the day they were made.

August 31st — All is peaceful here. The Turks are some distance away. Our wandering patrols miles away out in front have only come across nomad Bedouins. We are being issued with good food, bread, bully-beef, occasionally Maconochie rations, potatoes and onions, jam, tea and sugar, occasionally coffee and bacon — far and away the best rations we have yet had. The Heads must have dug up all the camels of Egypt to supply us. Each section does its own cooking, and these old palms smell some tasty dishes. We cook in empty meat and biscuit tins, and the ever handy quart-pot. Those of us not lucky enough to own knives use sticks. None of us have blankets (except the horse saddle-blanket), few, from the colonel down, possess tunics. We fight as we live and are a pretty tough crowd, thoroughly efficient, terrible shots both with rifle and machine-gun, ever alert and up to every dodge of both trench and desert warfare. At any time, at less than five minutes' notice the regiment can be in the saddle prepared to fight immediately, or to ride forty miles and fight at the end of it, and fight hard. Just as well we are trained to the knocker, because the Turk is a past-master at this type of warfare. Efficiency means our lives. …The regiment received 100 pounds worth of bucksheesh tinned fruit. Great! Good luck to whoever sent it.

September 8th — El Fatia. Have camped in this oasis for nine days. Times are easy, have been out only on one twenty-four hour patrol. Other squadrons' patrols have smashed a few Jacko patrols — tiny fights in the desert, interesting only to those who individually take and give the bullets.

September 11th — The Victoria Racing Club have sent us a splendid gift of tinned fruits, lollies, first-class tobacco and some shirts marked "from the Mayoress of Melbourne." Everything was appreciated immensely.

September 15th — To-night we are going out on a twenty-five mile stunt, to attack the Turkish advance garrison at Mazar, which is forty-four miles east of Romani. (Their base is at El Arish, miles farther back.) We attack at dawn What an agonizing trip in the sand-carts and cacolets the poor devils of wounded will have!

September 16th—We left Fatia this morning at 2 a.m., arriving at this oasis, Geeila, at dawn. Wells have been dug. A taube just buzzed overhead but got the shock of her life when four 'planes instantly rushed her. She fired a startled burst from her machine-gun and fled. I'm afraid our chance of surprising Mazar is gone.

3 p.m.—We hear now that when the taube fired she killed a man and horse of the 10th Light Horse. Unlucky poor chap We are to move out at sunset, cover the long dry stage, and attack at dawn, first surprising a chain of entrenched outposts placed a mile out fronting their redoubts. We have to gallop them down.

September 18th—Arrived back here at six yesterday morning, dead beat, a truly awful trip. Here goes. At sunset we rode straight out into the desert. Night came in utter silence. The 3rd Brigade was riding across to our right, somewhere, to attack Mazar from the east. The Camel Brigade, with guns of the Hong Kong and Singapore Mountain Battery, was mooching across the desert to destroy a Turkish post at Kasseiha; after which they were to get right behind the threatened position. We believed the 1st Brigade was somewhere away to the left. Our brigade had a squadron of En Zed machine-gunners. The travelling was in among and up and over steep sandhills all night, many of the gullies were pitch black and precipitous, a man didn't know into what pit he was falling should his horse roll down. "No smoking!" "No talking!" The old 5th was the advance guard for the 2nd Brigade, our own troop riding on the extreme left flank screen, so we kept our ears and eyes wide open; that is, as much as we could; most of us had already been two nights without sleep. Presently the moon came and made all the desert silver except the hillsides and donga gullies which gaped blacker than before. We in the screen, responsible to guard the main column against surprise, strained our eyes at the bushes that looked so like men. We never knew what second might bring a storm of bullets from some outpost or redoubt. Our orders were that if we rode on an outpost to immediately open fire and let no man escape. So we rode on, careful to keep the shadowy horsemen to right and left of us always in sight, and to keep with the nearest visible spider lines of men that stretched back in touch with that indistinct, long black shadow winding low down in the hills. Occasionally the column would halt, we would know by a warning "hiss!" and sight of the shadow horsemen behind halting, and we'd tumble off our horses, but keep staring to front or flank. Presently there came a very unusual halt, almost two hours. The Heads were giving the Camel Corps time to detour away around Mazar and cut the Turkish communication wires with El Arish.

Presently, we spied the shadows moving up on us, all magnified, all silent. We leapt on our horses and moved off again. At long last there came steel in the east, and presently a pink glow. Thank God for the dawn. Once the sun was up, it would drive the sleep from our eyes, not the sleep but the intense craving for it. We strained eyes and ears, for we must be almost on the Turkish outposts. Would the rifles never crack! Suddenly we half wheeled our horses and slung up the rifles as a patrol of camelmen dashed from the bushes urging their monstrous steeds with hoarse, low cries. But a troop galloped between them and Mazar and laughingly cut them off.

But the Turks were aroused. Men jumped up from the bushes ahead like wallabies and mounting fast camels were away for their redoubts. Crack! Crack! Crack! Crack! The game had started. Even the horses knew!

We pressed on, expecting the bullets. Soon the regiment galloped into action, one squadron within two hundred yards of the Turks. We all thought that now would come one great gallop of the brigades down the hill and into Mazar. But the order never came.

"Bang! bang! crash! crash!" sang the anti-aircraft as they slung their shells at our circling 'planes.

"Bang! bang! bang! bang!" answered the Somerset Battery.

"Rut-tut-tut-tuttuttuttuttuttut," came the machine-guns' throaty chorus.

C Troop was ordered to canter north in search of some strong outpost that threatened our flank. We had to keep them occupied lest they be reinforced by machine-guns.

This little "private" stunt kept C Troop out of the fight, as it happened. We cantered away as the sun bounced out of the hills and lit up the stunted bushes. Presently we cantered on a cunning little redoubt, so craftily winding in and out among the bushes that it was impossible to see it until we rode right on it. If it had been held by determined men, it would probably have wiped us out! But the Turks had fled back to their main positions. A coarse desert grass had been packed into the trench sides to prevent the sand filling in. We admired its neat and businesslike air, and the cunning way in which the apparently growing grass hid the lips of the tiny trenches.

We rode warily on, the fight flaring up and down away to our right and in the air, but around us was only the occasional hum of long-range bullets, everywhere about us low sandhills densely covered with scrub— and the expectancy of sudden death everywhere.

We found quite a number of outpost trenches, but the Turks had all cleared away back to their big redoubts in Mazar. We turned to rejoin the regiment, climbed a hill and gazed down on Mazar, and to our intense surprise saw a smoke-puff, then flash of flame, then bang! bang! bang! bang! crash! crash! crash! crash! We were actually gazing at a Turkish anti-aircraft battery in action. We turned and galloped for the regiment and the guns. What a wonderful target! We had a fairly long ride and wondered at the peculiar firing: it would almost die down then roll out to a long hoarse growl, to die down again to regimental firing, only to break with a sudden intensity and ripple distantly away.

We trotted up to our led horses but to our intense surprise were told not to go into the firing-line. We saw General Chauvel and the Old Brig, earnestly discussing the situation. We wondered what on earth had happened.

Across the desert where the 3rd Brigade was, the firing broke out like the wind-swept roar of a bushfire. Brigadier Royston sent word that he could take his position, but at the cost of an awful lot of men as the enemy in front were unexpectedly numerous and in strongly entrenched positions. Also, it was not certain that the Camel Corps had succeeded in getting to the rear of Mazar.

An Australian Light Horse regiment crossing a pontoon bridge in Palestine.

28

We learned afterwards that Chauvel had declined to sacrifice men unnecessarily. Our horses, too, after an exceptionally rough thirty-mile trip had not had a drink for twenty hours, and must return another twenty-five miles before they got a drink.

We retired at midday, furious about it all, certain that determined gallop would have ridden down the redoubts. I'm blessed if the efficient Turks didn't farewell our going by converting their anti-aircraft guns into field artillery and shell us.

By Jove, I was thirsty! The heat from the sand rose up to a man's face. Presently the rattling fire of the rear-guard dimmed behind us, while the brigades concentrated in a valley and caught up to the ambulance column with its broad-wheeled sand-carts and big red cross, its lurching camels with a stretcher strapped on either side, swaying like a ship at sea. There were mules and horses each dragging a light wooden sled on which was strapped a dead or wounded man. He would have an easier ride than the cacolet men. But at every jolt of the sled—

I felt awfully glad our casualties were practically nil. I pitied our odd unrecoverable dead lying away back. They must be very lonely. Our wounded were in hell, but at long last they would reach a hospital and sleep, sleep, sleep!

In the afternoon we saw disappearing over a rise ahead, the rumps of the Camel Corps. Our horses, with heads down and chests breathing deeply, knew there would be water at the end of this. Soon we caught up to the cameleers. I looked back. Away behind was the slow-travelling Red Cross column; then the artillery with the horses straining at the guns; then far-reaching regiments of Light Horse; horses, horses, horsemen everywhere; all Australian horsemen with the exception of the En Zed machine-gunners and the Tommy batteries; sunburned, hard-faced, laughing, joking, growling, cursing, smoking, thirsty rough men on rough horses on rough work. And the Camel Corps of Australians, En Zeds and English stretching back across the sombre hills.

We swore at the Heads, wherever they might he. With this little lot we should have eaten Mazar. So we spoke our thoughts.

Perhaps the Heads knew that the dice were loaded.

After another nine miles we were surprised to see a long convoy of camels coming towards us between the hills. I don't know whether the horses sighted the phantasies or smelled the water within them, but a faint ripple of neighing, seldom heard now, broke out down the column. Horses threw up their heads, open mouthed, sniffing eagerly. They pressed on and the sand sobbed the depth of their desire.

We met the convoy—the horses went mad—they rushed it—at sight of water we could not hold them—they swarmed like mad things, pawing, panting, jostling, straining. Two of us held back the section horses while the other two vied around the phantasies for water, but immediately we got our buckets full all horses rushed us. A dozen gasping mouths into one bucket—struggling animals, shouting men, rattling of stirrup-irons, pressure of horses' bodies, spilled water—open-mouthed men trying to catch the splashes from the buckets—plunging circle after circle around each phantasy, horse-holders with straining arms finally dragged over in the sand. It soon ended, but the horses struggled to lick the wet sand, frantic. There was not sufficient water—not even a squadron in our regiment got a drink! And the regiments coming behind—You see, there had been other regiments in front.

The regiment pushed rapidly on and presently passed all the Camel Corps, making for water, water, water! How my bones ached! I thanked Christ when the sun went down. Eventually we saw lights among the black palms of Salmana. The horses were frantic—they couldn't go faster than they did. Within the oasis, spelling troops had filled the water-troughs—the horses rushed these troughs, their heads in rows went down, stayed down; we could not drag them away. They felt like the weight of elephants. They would not come away. The water was brackish too.

We gave the horses the tiny bit of grain we had left, then linked them up. There were no rations for us, no fresh water. We fell on the sand asleep—except the unfortunates on picket duty.

At 2 a.m. we were aroused, light rations issued, but no water. Straight-way we saddled up, joined the brigade, then the black mass of men rode across the eternal desert to Hassaniya, where we are now.

September 19th—Why the hell don't I get any letters or parcels? My mates are always getting them: I don't get a lousy newspaper, even.

...The regiment is short of matches. It is amusing when it wants to light its pipe. Some lucky beggar, with studied ceremony, pulls out a precious "strike." The cry goes up: "Match a-light-oh!" and the pipe smokers rush up with spills of paper to thrust in the flame. You ought to hear the swearing when their haste puts the match out! The cigarette smokers light up from the bowls of the pipes ... No doubt we are a queer lot, a scatter-brained, joking, swearing, laughing lot. Last night, the whole crowd were trying to sing comic songs. They made the oasis hideous with choruses of the most idiotic songs I've ever heard—and only a few hours in from a stunt which for sheer physical weariness it is impossible to describe!

September 27th—I wish the war was ended.

September 28th—The Turks have evacuated Bir-el-Mazar, fearful that we may complete the job next time.

...The V.R.C. have again sent us a huge parcel of tinned fruit stuffs and first-class tobacco. May their shadow never grow less!

...To-morrow we leave El Fatia for Kantara, for a "spell." Apparently the fighting is over—for the time being. The engineers have been following us up by building a railway line, also a pipe-line in which water is pumped all the way across the desert, from Kantara Canal. The infantry are coming as the railway line advances, along the "wire road." The sand is scraped flat, wire netting is stretched upon it, and pegged down. The idea came from the Queensland bushmen used to putting down wire-netting across the sandy river-bed to act as a road.

Building a railway and big water-main across the desert, under taube attacks and enemy activities, is the sort of job the pyramid builders might have taken on.

Infantry and Yeomanry are to hold the positions we have won. What will our next move be? Who knows! who cares! But at last we might get a little spell from the long marches, the thirst, the sleepless nights and all the weary rest of it. Anyway, we are glad to have lived through the campaign. It has been intensely interesting and exciting; though the Turks do call us the "Mad Bushmans."

God knows there should be no heartache at leaving the desert, but this afternoon we were making some little wooden crosses.

October 1st—Romani. We rode from El Fatia at dawn yesterday—a lovely morning. The desert looked strangely "soft." We were laughing and singing as we rode over those eternal sandhills whose flimsy sands have lasted while the many armies that have trod them have blown away in dust. We reached the strange steel ribbons now creeping across the desert. A battalion of Tommy infantry tumbled out of their bush gunyahs to gaze at us. They appeared thunderstruck by this 'ragtime army'; many stared suspiciously. Certainly we looked the newspaper idea of a Turk. Some were dressed in riding-breeches, some in shorts; most wore leggings, others puttees. All were in sleeveless flannels, shirts, or singlets, few among the officers even possessed tunics. Some of us were shaven, others not. All wore dilapidated hats, most still flaunting a tuft of the once proud emu feather or wallaby fur, and often ventilated by a bullet-hole. A rough-looking lot. Before midday we sighted the shimmering bare hills of

Romani and soon were winding in among the redoubts. Desperate looking positions to take, the infantry had had three thousand five hundred Egyptian labourers to help prepare them. The infantry are wild that old Jacko did not tackle them. We watered at ancient wells reputed to have been sunk thousands of years before Bible history, then rode on to here and—surprise! As far as the eye can see are military camps, mostly Tommy and Scotch infantry. Where on earth have they all come from, and when? We are *en route* to Kantara in a few days.

…Last night was comical and lively. Fancy beer in the desert! plenty of it! The desert is getting civilized. When the whole brigade arrived—fifteen hundred men with a desert thirst—!

A vast hum of voices arose from the canteen. Bert, Morry, Nix and I took our quart-pots to see what was doing. Bert of course had been there earlier in the afternoon. He was happy and laughing—he even sang. The canteen was a huge marquee tent. When we fought our way inside, the roar was deafening; it was an atmosphere to crack a joke in, if only a man could make himself heard. Odd groups were gambling at "Crown and Anchor," a box and shaded candle seemingly their equipment. Each man's quart-pot was lathered in beer, each was firmly holding his own.

We pushed towards the perspiring barmen and big Morry grinned gloatingly. I noticed that the barmen's perspiration trickled into the beer, but no one cared—we'd drunk far worse than that. We got our quarts filled at last, then looked about for somewhere to stand. A man had to be jolly careful walking among those rowdy groups in the dark lest he spill his beer down another chap's neck and lose it—There were no lights for fear of a bomb raid.

They attempted to put us out, at last, then partly collapsed the tent on us. We eventually crawled out on to the cool, wondering desert. Nix pointed a wobbly hand to the sky in general and stuttered something about following "th' Star a Bethlehem" home, so I thought it safer to follow Morry.

We made camp somehow, and after a lot of singing and nonsense found our saddle-blankets and settled down to sleep, Bertie insisting on taking his trousers off as we were "away" from the desert now. Suddenly there came a wail of screeches accompanied by thumpings on a drum. There was no doubt about it, there were the bagpipes playing a rollicking march tune! There was a drum and yells and shouts and pandemonium coming through the distance! Up jumped Bertie, chucked his hands above his head and somersaulted, let out an awful screech, and howling something about his "Scotch blood stirring" sped away towards the advancing pipers. Nix and Morry and Stan and I followed and met the pipers and drummers coming down the track.

A rollicking crowd was following them up: the pipers piped the silly beggars home, Bertie yelling as he did a clan dance in advance.

We got to sleep at last. I think the Scotties chased Bertie home—I'm not quite sure.

General Chauvel, 1st Light Horse.

29

October 2nd—Romani. A glorious morning—no "Stand to," the first omission since—? has a man got to reckon in years now? I slept until five o'clock.

October 3rd—Yesterday, General Chauvel gave us a little heart talk on women, and the everlasting saluting. The women part was interesting, though he didn't tell us much that we did not already know. The saluting part, however, we listened to in unappreciative silence. The English Heads have been rubbing it into the general, and he turns around and lathers us. Then in a comradely sort of way he brought our visions of a spell crashing. It appears that our horses are so worn out as to be absolutely useless unless they are spelled. If nothing intervenes, small batches of men will be sent to Sidi Birsch, Alexandria, on short leave. So that's that! Good job the general told us straight out, though.

October 6th—Dueidar. What a change in the old camp since that long gone day when we galloped up to the hard-pressed garrison of Scots Fusiliers. Now are a circle of splendid redoubts, rows of tents shelter the 6th Fusiliers. Everything is transformed. Even we have tents. and there are actually regimental cooking ovens.

We prefer neither. We would rather cook and camp in our own little sections. I believe we are getting domesticated, whatever that may mean.

…Numerous fatigue parties—plenty of night duties. There is talk of three men out of each hundred and twenty being sent to Port Said for forty-eight hours leave. Some of the poor beggars have been two years in the desert without a spell.

The garrison here have unearthed a large Roman well, the bricks and mortar in perfect condition. It is about ten feet wide, circular, dome-shaped at the top. Running along it is a strip of cement about five feet wide and sixty long, like a cricket pitch; we wonder what it was for. The Tommies unearthed old coins. Our pumps water the horses from that well now. Those tough Roman centurions did their work to last. The desert may cover it up, but—

After all, I believe memory is stronger than the desert; it will last longer than the pyramids. Two thousand years ago, the child Christ passed along this road when Herod drove His parents into Egypt. Well Christ has gone, but His memory is here. A queerly personal thought that Australian soldiers have actually trodden in the footsteps of Christ!

October 9th—The age of miracles is here in the desert!—I am in Port

Said. Said the sergeant: "Idriess, you and Crewe are to report to the Rest Camp in Port Said. Two men out of each troop, who are absolutely run down, go to the Rest Camp for six days spell."

I thought it some silly joke, until they paid Crewe and me £4. Surely the war is over!

October 10th—We visited the Armenian refugees' camp to-day, four thousand men, women and children in a tent town. The women are industrious, working at needle and basketwork, the old men fashioning combs, spoons, knives, etc., out of bone, wood, or any old thing at all. They sit on the floor at their work. The girls are plump, dark-eyed, dark-skinned. Pitiful tales are many. Orphans, mothers with half their children missing, fathers whose wives have been massacred by the Turks. But, it is hard to have pity for them. Their young men will not work, let alone fight. People who are too mean to fight for their loved ones, their homes and their own lives, must take what they get in these civilized days. In only isolated cases did the Armenians put up a fight against their taskmasters. The Turks are credited with having slaughtered a million of them.

October 25th—Our holiday is long since over. We were allowed five heavenly days. The Rest Camp at Port Said was a military institution run solely for the soldiers' comfort. We were intensely surprised. Plenty of leave—no guards—no distasteful restrictions. And the men did not abuse their liberty. The City military police were gentlemen—they left us severely alone and we reciprocated. We had such an awfully nice time that I can't write about it. You see, the desert—

November 16th—We've had a fairly easy time. The old regiment keeps in training. We've easily won all the rifle and revolver matches against the local garrisons. All have had their five days leave. Gus Gaunt is back, fat as a butcher's pup. There are black soldiers here now, B.W.I. (British West India) men from Jamaica.

November 22nd—We move desert-wards again soon. The nights are damned cold.

December 2nd—The Gloucester Yeomanry took over Dueidar. We are once again on the track of the Turk. Camped at Batar. The trip has been a picnic with, glory of glories, "Stand to arms" at five o'clock! until we catch up with the troops ahead. The 6th Light Horse were bombed at Bayud.

December 7th—Our patrols clashed with Turkish cavalry patrols yesterday. There are intriguing rumours of crack divisions of Turkish cavalry coming up, to wipe out the "Mad Bushmens."

December 8th—Our brigade has to watch the right flank against the Turks on the Maghdarba Hills. The other brigades are pushing on towards El Arish.

December 14th — A squadron is right away out here at Gamal oasis. We protect ourselves by a ring of outposts. Our section manned No. 1 post last night, a creepy position on a castellated sand-mound crowned by stunted bush. At 3 a.m. I was lying flat on the mound, peering between the twigs at the desert shadows. Morry and Bert and Stan lay shrouded in their greatcoats, sleeping deeply. To our right looked a precipitous sandhill with a gully running between us and it. What if the Turks were creeping up that gully to attack the camp? Fronting us were limitless sand-mounds, running right up to my rifle muzzle, splendid cover for stealthy Bedouin. To our left among the grey bush patches was a deep gully completely encircling the sand-mound. The Turks might crawl up it, then with one rush bayonet the sentry and his sleeping mates.

I stared out to the front, then slowly to the right, then back to the front, then to the left and gripped the trigger as my head turned round, round, until I was glaring directly behind towards that gully, then with a sigh of relief gazed out on the front again, and before the two shifts of two hours each was up was well kinked in the neck and nerves.

Our squadron is posted right where the Turks must meet it should they venture a surprise flank attack, and we are responsible to protect the brigades against surprise. I was glad when the moon rose and bathed the desert in silver. Its countless bushes all wore silver coats; everything stood out vividly.

We are "desert wise," so dare not gaze at any one bush for long. If you do, just as surely it will move. You stare — slowly it takes the shape of a man. You breathe deep — trembling! Will you wake your mates from their precious sleep, or will — the figure takes a stealthy step forward. You kick madly; your mates spring up shivering with bayoneted rifles thrust out to meet what is coming and — it is a bush!

Ah no, you never gaze at a bush too long! If uncertain, just glance away then look quickly back — and you can distinguish the very leaves of the thing if it be close.

But it is the ear, more than the eye, that whispers the desert's secrets at night. And last night, my ears nearly made my heart burst. Click, click, click, click of rifle-bolts from the bushes to my left. Too late to do anything — just lie with stiffened hair awaiting the volley — Then in a flash I remembered the Lewis Gun section away up on the hill to the right. Curse him; oh curse that fool gunner trying the mechanism of his gun! Stupid reinforcements! Fool! If the Turks had been around us those tell-tale clicks would have warned them!

The morning stars are startlingly bright. I watched one rise; it glowed before it tipped the sandhill top to rest definitely in a rosy twinkle — I could swear it was a light. Then it slowly rose a few feet, diamond bright—I could swear it was a Morse signal flare—But damn all outposts!

December 17th—"Stand to." Sun rising. On outpost again: thank heaven that night is over. It was so queer, though, I must write of last night while it is fresh in my memory. Soon the "All's clear" patrols will come in and we outposts will pick up our gear and trudge down into the oasis for breakfast. But last night—

I was sentry, staring out there, fighting to keep my mind occupied and awake and alert. I don't know why, but I began thinking of all the ancient armies, wondering even what other outpost squadrons have camped in this very oasis. I started musing about those old Phoenicians, and the Babylonians, and the kings of Persia and the Syrian cohorts, of the terrible "skin-'em-alive" horsemen, of the disciplined Roman legions, the Saracens and Crusaders and Arab hordes. I wondered if the Moslems and Christians and Mohammedans and Jews and idolaters and all the others mix together now! And I wondered if the phantom soldiers watch over our battles and if they gather in curious groups around the dying Anzacs. And I could see quite plainly the spirits of the Anzacs arising and staring at the weird soldiers gazing so silently back. And I could have sworn I heard, like a clear whisper, the long-drawn gasp of an Anzac: "Who th'-hell-are-you?"

I wondered if the phantom soldiers understood those Anzac swear words; I was sure they listened attentively to their tales of repulse and victory. I was positive of the attention on their faces and the cocksureness of the Anzacs as they explained the mechanism of the machine-gun, and artillery and flying-machines. And, queerest of all, I seemed to hear the Aussies boasting of gum-trees, and the swag, and of old Australia.

So real did the whole business grow that I know I forgot my duty—clean forgot everything except a shadowy sense of desert and real, live phantoms all around me. There was a calm, steady sort of chap standing beside me explaining, not in rough speech, but putting pictures in my mind and before my eyes lucidly and clearly; showing me the huge armies who are really watching our tiny army all along the old Darb el Sultani right from Egypt into Palestine.

I was entranced. He called me back into the present himself—turned with such a scowl on his handsome, swarthy face—I had just time to wheel around and whisper a challenge to an officer visiting the outposts.

30

December 22nd—With the regiment again Christmas billies have arrived, one to each man. Good-oh … A fat parcel came for me—to the section's delight.

…A taube (they terrify the Gyppo labourers) just dropped a message at Railhead: "Keep on with the railway line. We are ready to take it over any day now."… The Tommy Yeomanry general insists that the Yeomanry go out in front at the next big battle. The Anzacs have proved themselves: now the Yeomanry want a chance. Good luck to them.

December 23rd—When on desert patrol, we sometimes ride across dead Turks. The wind has eddied the sand away and part of the half-dried chaps gaze up. Most of our own boys have long since been decently buried. Their neat white crosses are dotted over the desert.

Afternoon—Well I'm blest—our troops have occupied El Arish. The Turks evacuated it—we thought hell would be played there. Meagre details are that the 1st Light Horse Brigade, the N.Z.M.R. Brigade, the 3rd Light Horse Brigade and the Imperial Camel Brigade, the Leicester, Somerset, and Inverness batteries, R.H.A. and the Singapore and Hong Kong Mountain batteries, all under Chauvel, concentrated on the Old Caravan Road at night and marched the thirty miles to El Arish. But the Turks had just gone.

December 25th—We are staggered. The old Desert Column has left the 2nd Brigade out of a fight—the first since we rode into the desert. A portion of the El Arish garrison stopped at Maghdaba, a fortified position thirty miles farther east. They were bombed there by ten Australian 'planes. Chauvel and his crowd followed them straight out from El Arish and dismounted within four miles of their bright campfires. Jacko never dreamed that the Desert Column, after a thirty-mile night ride, could possibly carry on with the business and ride out another thirty miles. It was an epic fight. The position was entirely surrounded. The Turks fought stubbornly all day, our fellows getting closer and closer until just at sunset they fixed bayonets and regiment after regiment charged the redoubts. Only one crowd of Turks stood up to the steel, but the other redoubts fired right to the moment that the Anzacs jumped down into the trenches.

The Turkish losses were immeasurably heavier than ours although our fellows were attacking from the open desert and the Turks were snug in deep trenches supported machine-guns. All the garrison were captured with their Krupps. Our wounded were collected in the dark, little fires were lit beside them so that they could be picked up by the stretcher-bearers groping across the desert. Then came a dreadful march back, for columns of Turkish reinforcements from Shellal were hurrying Maghdaba. The wounded had a fearful time in the dreaded cacolets.

Later—A very peculiar story is being discussed throughout the Desert Column. It appears that the troops when riding back the thirty miles from Maghdaba were enveloped in blinding clouds of dust. Nearly the whole column was riding in snatches of sleep; no one had slept for four nights and they had ridden ninety miles.

Hundreds of men saw the queerest visions—weird looking soldiers were riding beside them, many were mounted on strange animals. Hordes walked right amongst the horses making not the slightest sound. The column rode through towns with lights gleaming from the shuttered windows of quaint buildings. The country was all waving green fields and trees and flower gardens. Numbers of the men are speaking of what they saw in a most interesting, queer way. There were tall stone temples with marble pillars and swinging oil lamps—our fellows could smell the incense—and white mosques with stately minarets.

It is strange to hear the chaps discussing what they saw, as they sit smoking under the palms. I don't think they would talk so openly had it not been for a general riding with his staff. Suddenly he and a companion officer galloped off into the darkness. It has just come out that both officers suddenly saw a fox and galloped after it!

The men now don't mind telling what they saw, for when two of our Heads saw strange things, well—

The En Zed Brigade arrived back at El Arish at six in the morning. Poor beggars, they were immediately heavily bombed by 'planes, the morning of Christmas Eve!

And this is Christmas Day! Great Caesar! Which reminds me: we are very close now to the land which gave Christmas to the world.

...The Navy has appeared off the El Arish coast. Minesweepers are busy. Two 1st Brigade chaps, rushing to the shore for a swim, were blown to pieces by a mine. A thumb was the biggest part of them found.

December 26th—General Chetwode (he made his name at Mons) addressed the troops who fought at Maghdaba, saying that the mounted men at Maghdaba had done what he had never known cavalry do in the history of war: they had not only located and surrounded the enemy's position, but had got down to it as infantry and carried fortified positions at the point of the bayonet.

Which was a nice compliment, coming from an English general who was sent out from France especially to boss the show. Certainly it looks as if Maghdaba is one more milepost towards the end of the war.

Maghdaba was full of thrill. When redoubt after redoubt was falling to the bayonet the troops got wildly excited Two Westralian troops of only forty men charged a redoubt of four hundred Turks. They galloped yelling straight over the trenches, but seeing annihilation galloped on. Lieutenant Martin's horse was killed but his cobber lieutenant and Sergeant Gynne galloped back and got the lieutenant safely away. Soon afterwards a squadron of the 2nd Light Horse galloped shouting at the same redoubt, leaping over the dead and wounded nine of the 10th Light Horse and pounding into the trenches with their horses, firing through the dust from the saddles. They shot up the Turks and took the redoubt though the Turks were four times their number and were supported by German machine-gunners.

Royston's men are chuckling over their prized brigadier. In the big charge where the 10th captured seven hundred prisoners Royston as usual was galloping all over the battlefield and suddenly found five Turks peeping from a trench with levelled rifles. The brigadier instantly raised his cane and shouted furiously at them in Zulu, and I'm blessed if the puzzled Turks didn't drop their rifles and hold up their hands.

Khadir Bey was captured in No. 1 redoubt; its garrison only surrendered when the Australians and cameleers got into them with the steel. The 8th and 9th Light Horse, after suffering under intense fire, charged furiously into one redoubt that fought as furiously back, steel against steel. It must have been a mad few moments; the chaps get wildly excited as they tell us about it.

When all the charging regiments met together in the grand rush that overwhelmed the central redoubts the scene in the gloaming was of the wildest confusion of mixed Anzacs, sweat-covered and hilariously laughing, of excitement-maddened horses, and sullen Turks. The regiment is wild at having been out of it all, even though all the fighting regiments had been four successive nights without sleep.

…The section was on outpost last night, with a thunder-storm. Yes, a thunder-storm! The lightning lit up the desert with flaming fire. The smallest bush was distinct—even the grains of sand. The thunder rolled and crashed—then eerie silence. I crouched among the bushes, shivering under the greatcoat, waiting for the rain and—the Turks. Thank heavens, the Turks did not come. Both evils on the one night would have been as much as a lonely outpost could bear.

…It's been miserable ever since Christmas Day—storms and icy winds. The rain is not our rain, there are no puddle-holes and good old sloppiness. The desert simply swallows it up and waits for more. Some rain hisses as it falls, but generally it just drops silently, not much of it but miserably told. Under cover of night we leave the oasis and steal away to our outpost positions, each section going in a different direction. With our hands we dig holes in the sand, six feet long and two deep—ominous. We cover the wet sand with bushes, wrap our saddle-blankets around us (we must not unsaddle the horses), get into the holes and sleep until it is our turn for duty. It is quite cosy in the sand-holes, the wind blows over the top of us and the blankets shield us partly from the piercing rain. But when a man's blanket is soaked, he is miserable. Some outposts dig a big enough hole for two or three men to sleep in, and share the blankets. It is much warmer and takes the rain longer to work through the doubled blankets. A wind howls all day and night, the clouds gather like a creeping black wall that steals to the very ground—then comes bitter rain. Sentries on duty stare across the dessert, their fingers numb on their rifles.

…We hear that the troops away over on the coast El Arish have experienced bitter gales. A trawler was totally wrecked. The vessels landing stores are having a wretched time. The big wharf erected by the Australian bridging team has been engulfed by sand thrown up by the seas.

1917

January 9th—At last we have had our scouting trip. About four days ago Major Bolingbroke sent for me:

"Still got that old scouting idea, Idriess?"

"Yes," I answered, lively.

Then the major explained that the Bedouins who spy on our observation posts in the dawn, are believed to come by night from the ranges S.E. of us, but no one knows where exactly. It was my section's duty to go out and if possible cut the tracks of the patrols and follow them towards their camp. If lucky, we might capture a Bedouin or two and see what's doing. But we must take the greatest care not to be cut off and trapped.

The first proposed objective was a hill named El Risha, nineteen miles from camp. No one had ever been out there, and 'planes could see no oasis; so apparently there was no water. How the Bedouins obtained it no one knew.

The major loaned me map and compass and protractor, and together with Captain Patrick we marked out a probable course.

Old Morry and Stan and long Dan Jones were delighted. Bert would be mad—he was away with the regiment on a duty.

We rode off at 3 p.m. and after six miles got a glimpse of the big range. At sunset, we found the compass pointing direct towards a whopper hill apparently sixteen miles away. As the range faded into darkness we pressed on by sense of bushmanship alone.

It promised to be an interesting patrol. We four were in the immemorial desert of the Bedouin, the treacherous Amalekite of the Bible, and with every step we were going farther in, now utterly isolated, beyond any hope of succour. It was a bit thrilling.

The rain had ceased, the night was clear and bitterly cold. A full moon shone out transforming the black sandhills into silver-grey ones. The twigs on the bushes were like silver foil. Presently we stared down at the tracks of a day-old camel patrol, the imprints on the sand just as plain as the moon was in the sky; we looked at one another. The sand-dunes became higher and steeper as we steadily climbed. The horses were cold and fresh and travelled very fast. After each hour's riding we gave them a quarter-hour spell as we must keep them fresh, no matter how far we travelled, in case we were chased. Twice we rode on old Bedouin encampments, shadow camps in the silence.

At about nine o'clock we cut the first fresh tracks, five were camels and two were naked feet. Our eyes, and above all our ears, were keyed to the highest pitch. We could see a surprisingly long way, even to the black gorges now gaping between the hills. And our ears could *hear* the silence.

Half an hour later we gazed down over our horses' necks at the tracks of many men. In the intense moonlight we could see all glistening the very grains of sand their toes had thrown up. Those tracks were only hours old.

By now we must be within striking distance of El Risha. I decided to camp, as at any moment we might ride into a Bedouin encampment. To push on might mean that they would be behind us at dawn, and then good-bye to our getaway.

We climbed far up the side of a gloomy hill until we found a sheer wall. With our backs against this, no possible tribesmen could either surround or rush us by surprise. One man held the horses and watched, while the other three lay at the horses' hooves and tried to sleep. The horses were uneasy, snuffling and staring at bushes. Before the east brightened we fed the horses and ourselves. In the dark I climbed the hill-crest and at the greying of the clouds peered over the skyline. There, not three miles away, was the sombre black crest of El Risha. I was quite pleased that through the night we had ridden in a direct line towards it.

We mounted and moved ahead very cautiously. Tracks were plentiful. On a hill-crest in black silhouette was a roughly constructed outpost shelter. We cantered straight at it—it was empty. Suddenly Stan called: "Look! camels!" and there gazing at us were two fine big animals. We spread out immediately and galloped straight for them, thinking them the camels of a Bedouin patrol. Up jumped four more camels and the six sped at an amazing rate for El Risha. They looked like gigantic emus. It was no use chasing them.

The light grew momentarily brighter. We felt quite happy at discovering there were rocks about; we had forgotten what a rock looked like. Our horses became quite prancy upon the harder ground. Suddenly we found ourselves right among a mob of camels that were coming down from El Risha and, although we did not realize it then in the half light, being driven by Bedouins down on to pasture grounds.

Suddenly Dan called hoarsely: "Look, a man!" and there among the bushes and rocks was a Bedouin half crouching, half running. We spurred towards him, and the ground seemed to swallow him up. Our horses were snorting as we encircled the crest and with rifles at the ready flushed the bushes. Seeing the game was up, he jumped from cover and came down towards us pulling a paper from a rude goatskin wallet. He was dressed in the rough costume of the outback desert Bedouin. The paper was a Turkish proclamation of some sort or another.

We thought with the assistance of this man that we might be able to muster the camels and drive them off. The beasts were now browsing on the hillsides and down the steep gullies. With the rising sun we saw that coarse grass grew amongst the bushes and rocks. We were in the shadows of rocky hills that frowned all around El Risha. Stan set off on a half-mile canter to round up the head of the camel mob, while we spread out and mustered them in. They were actively afraid of the horses, but Stan soon had them turned, and we laughed as he came tearing down the hill with the camels lurching before him, their long heads coming towards us like an ungainly mob of giraffes. The devils were as agile as mountain goats. We bunched them up, but camels were disappearing over the skyline in all directions. I became a bit anxious, for the Bedouins wouldn't stand that sort of thing for long and without doubt they had a large camp near El Risha. What saved us was that they did not know our strength, but soon they must know.

We hung on to sixty of the camels, but our Bedouin captive was sulky and would not work. The camels soon became unmanageable; one big bull wheeled and with a roar and slobbering tongue went straight back for El Risha, the mob thundering at his heels, Dan and Morry vainly endeavouring to head them off. Our horses got quite excited and were noticeably hostile to the camels. The beasts won.

Morry and Dan came back with another dishevelled Bedouin. We decided to quickly push on to El Risha, now only a mile away, and have a peep at the country over on the eastern side. We had no time to waste, and expected a hornet's nest to come buzzing out to round us up as we had tried to round up the camels. But we must get some definite information.

When close to the big hill Stan suddenly called: "Look behind!" And there, five hundred yards behind us, was a man running on the skyline. He dropped behind some bushes. This looked as if the Bedouins were already trying to cut us off. We decided it was best for Morry and Dan to stay there with the two Bedouins while Stan and I hurriedly climbed the hill. Morry and Dan chose a position where they could not be surprised, and could keep within rifle-shot of us.

Stan and I scaled big El Risha, nearly all rock. A superb panorama stretched before and on either hand to a far horizon. El Risha dropped precipitously to flat country below. Miles ahead were sheer rocky ranges. The flat was pimpled with abrupt little hills. To the south were encircling hills like walls of sand. To the nor'-east the sandhills were lower, merging into rocky ranges. More than a mile in front of us stood like a gigantic sphinx a flat rock a hundred and fifty feet high, six hundred long, by a hundred wide; and standing in its centre was a sentinel. We watched him for quite a time. He did not move. Stretching right across the flat towards the foot of the big ranges, were clearly visible the tracks of our late departed camels. Evidently the main encampment was farther away than we had thought. Stan and I determined to investigate the figure on the rock. Making sure that Dan and Morry were all right, we rode down the big hill and trotted across the flat towards it. It was a long way. I climbed the rock and found the damned sentinel to be a cairn of stones. Gazing along the camel-tracks, I saw eight men mounted on swift camels coming towards us. Behind them was the smoke of many fires. It was time for us to clear out.

When trotting back towards the big hill we saw a Bedouin tearing along a camel-pad. We galloped—he tried desperately to get away. He panted, and gabbled at us in a strange tongue. He was very unwilling to come back with us, but we had not time to argue. He had been spying on Morry and Dan and was running back to warn the Bedouins in camp. We lost no time, quickly crossed the big hill, and found Morry and Dan. Stan was determined to rope in at least some camels. He convinced the Bedouin, cantered away and presently drove a mob of sixteen towards us. We added another eight as we hurried along. Some of these were young.

For the first five miles we had a lively time driving them and expected every moment shots and a flying band of Bedouins at our heels. They contented themselves with dogging us from a distance, evidently afraid themselves of riding into a trap.

Presently the young camels got tired. The older ones then waited back for them and were consequently easier to manage. I had never thought we would be able to get them in, but as the miles went by and nothing happened I got so optimistic that we all started whistling. It was a jolly strenuous ride however, and it was with relief that we got nearer and nearer our camp.

We reached there at eight o'clock that night, camels dead tired, Bedouins dead tired, horses and ourselves dead tired. The squadron made quite a fuss of the capture.

Dawn capture of wandering Turkish patrols during the Beersheba fighting.

31

January—The camels and Bedouins were driven across to Brigade Headquarters. Eventually orders came from the English general away back in Kantara, to load the camels with food-stuffs and let the Bedouins go with a message to their people that the English are at peace with the Bedouin, and are their friends. Well, it sounds all right, but the troops want to know why the Bedouins cut the throats of our wounded, why they dig up our dead just to rob them of their clothes, why they signal the Turkish artillery of our whereabouts, why they come twenty miles even across the desert to spy on and snipe at our outposts. It is a queer idea of peace.

...The paper the Bedouin handed me was deciphered a Turkish proclamation informing the Bedouin that the Sherif of Mecca had revolted against the Turks, had taken the town of Medina, and gone on the side of the Christian dogs. The proclamation exhorted all true Mussulmans to fight against the Christians lest they overrun the land and kill all the Bedouin in the hills, etc.

...One of our Bedouin friends was wearing a Turkish military overcoat, with a bullet-hole through the sleeve.

...The whole Desert Column is excited. A Light horse patrol came in saying they had ridden over miles and miles of green grass. We don't believe it. We know no particulars, our squadron is isolated so many miles away from the troops ... This afternoon, we hear that the 1st Light Horse Brigade made a reconnaissance to the big village of Sheikh Zowaid, half-way to Rafa from El Arish. Rafa is on the border, the Turks have strong redoubts there. The men bring back strange tales of having seen a plain of grass. The 1st Brigade call their brigadier "Fighting Charlie". Anyway he bought a dozen hens from the Sheikh, and the hens lay eggs. We don't believe it.

...There's a mysterious, whispered-of craft follows us up along the coast. She is called the *Ben-my-chree*. She is a mother ship of sea-planes, Royal Navy fellows. They are air-raiders from the sea.

I'm blest if another scrap hasn't been fought, and we out of it again! On the 8th, the troops marched out on a thirty-mile night ride, to encircle Rafa on the border of Egypt and Palestine near the old Egyptian police post. Before daylight the troops surrounded Sheikh Zowaid, then carried on while the En Zeds rounded up the big Bedouin encampment at Karm Ibn Musleh, but not before the Arabs sent up flares and that wailing, warning cry of theirs. But it was too late. The troops had surrounded Rafa at dawn and a ten hours fight began.

They tell us Rafa was a tough nut to crack. It is dominated by a hill El Magruntein, which is tiered by trenches surveyed by German engineers. The other low hills around El Magruntein were all converted into redoubts. The country surrounding them was gently sloping with not even a bush for cover, there was only one tree visible and that was on top of a Turkish hill. The artillery opened out, then the troops galloped up close until their horses were being slaughtered, then dismounted and fought on foot, getting closer and closer all through the day, until at sunset they were within bayonet distance. The whole crowd charged and stormed redoubt after redoubt, then took El Magruntein with the steel almost in the dark.

The little Red Cross sand-carts went right up to the firing line in plain view of the Krupp guns. Transports galloped up too with ammunition. The only cover the troops had was our own rifle-fire. It appears that the brigades simply poured their lead into the Turkish redoubts thus keeping down the Turks' heads while our fellows crept closer and closer. The Turkish High Command away back in Beersheba sent out battalions of reinforcements from Wadi Sheikh Nuran Khan Yunus, and from Shellal, but our fellows held them off too and won out. We don't know particulars, but apparently it was a very close go.

After the fight they buried our dead, collected the wounded in the dark, then brought them and the prisoners and guns back to El Arish through the night.

…We hear that at Rafa, at daylight, when the Auckland regiment reached the boundary-line, the old colonel halted his men, rode on alone past a boundary pillar, took off his hat, and thanked God that he had at last been permitted to enter the Holy Land.

…The Battle of Rafa was fought half in Palestine, half in Sinai. But we are not the first. It is a battleground of ages past.

…They tell us that when they crossed the wadi (old river-bed) at El Arish, it was in brown flood, "spate" as the Arabs call it. We are beginning to think, what with grass and floods and hens' eggs, that we will soon be entering the land of "milk and honey."

…The Light Horse and En Zeds speak well of the Tommy guns at Rafa.

…The Rafa battle, like Maghdaba, was almost given up at sunset because of the lack of water. Chetwode, the new English general, anyway does not believe in staying scores of miles in the rear. He rode out to Rafa with Chauvel. The redoubts all fell to the bayonet when hope was given up. As the New Zealand Brigade was charging one redoubt the Cameleers charged another. Both brigades at the same moment saw the Turks in their trenches fixing their bayonets and the simultaneous roar swept all over the battlefield. The 1st Light Horse Brigade charged in the midst of the excitement and every Turkish trench was overrun.

The boys are enthusiastic about a patrol of six Ford cars armed with machine-guns that went into the fight with the Yeomanry. The old 5th are curious to see the new weapon.

February 1st—Geeila oasis. Have rejoined the regiment. The usual patrols. Jacko refuses to come out and face the Desert Column, but our 'planes report he is working strenuously on a twenty-mile line of trenches and wiring redoubts from Shellal to Beersheba. Also, he is converting the hills of Gaza into a chain of what the Germans boast will be impregnable fortresses. He will presently have a line of trenches right across Palestine. There'll be something doing if we have to break through. The whole army is wondering. Are we to invade Palestine? With the taking of Rafa, we have cleared Egypt and Sinai of the Turk. What next? Only the desert answers, in that utter silence which seems all whispers.

February 2nd—Regiment on the march. East, ever east. We have rejoined the brigade at Moseifig.

February 3rd—El Mazar. On outpost duty last night. It was fascinating to see the trains rushing past. How the old desert has changed since that awful ride of ours! But the desert has not changed; it is just that Britain has brought along the twentieth century into a land that was ancient when Christ was a child.

February 7th—Mazar has one solitary palm-tree, and the forlorn tomb of Abu Gilban, whoever he is. The Old Brig, said the tomb looks like the remains of an aged chimney. It does.

…Turkish cavalry patrols are reported away out on the right. Our Secret Service agents report that the Turks are hurrying up crack cavalry divisions to wipe out the Australians.

…The weather is exhilarating, but these silver nights bring us winged death from the air. The taubes swoop so low we can see the goggles on the airmen's shadows. However, they don't seem able to distinguish us if we don't move. If a man does, though, they drop their bombs, circle, and machine-gun the very heart of the camp.

The desert is trying to swallow the railway line. After each windstorm, gangs of Gyppos shovel the sand off the rails. Railway men tell us that moving sand sometimes stops the iron horse.

February 8th—On the march. Upon the Bardawil again.

February 10th—A whopper oasis close by the beach. The minaret of El

Arish is about three miles away. How strange it is, the old sea battering the desert shore, a train puffing along the beach. Our horses don't know what to make of the sea, and we ourselves were astounded this morning. A taube flew unusually high over El Arish. Bang! Bang! Bang! Bang! Crash! Crash! Crash! Crash! Puff-clouds floated around her. We just couldn't realize that we had anti-aircraft guns at last. Heavens!

February 12th—El Arish is a string of large oases about seven miles long, most oases hugging the shore. A sombre sandhill by the sea is topped by the domed tomb of Nebi Yesir. By a beautiful grove of palms is the spreading mouth of the wadi, once famous as the "River of Egypt." It only runs with brown water when storms burst in the far distant hills. Encircling these oases are bare sandhills. In the centre of the oases is land cultivated exactly as it was in the days of the Pharaohs. About two miles up the wadi, perched on sandhills, is the scattered town of El Arish, once famed to the ancients as Rhinocolura, a huge town lining both banks of the then river. To-day, conspicuous above the town is its mosque and minaret standing on a rise beside the ruins of a massive stone fort. This once defied Napoleon for a week. Its great walls were brought down by the guns of the British Navy. At a distance, the town is indistinct with the colour of the desert, but looks brighter when close, for the mud-brick houses, quaintly Eastern looking, are coated with a yellow plaster. People appear a fairly bright lot, for Arabs. Numbers of the girls are quite pretty, even though pregnant from the Turkish soldiery. We were surprised that some of the people had red hair. Red-headed Arabs! By Jove, we were curious. However, some learned officer reminded us that Napoleon's men had been there.

Across the wadi are other magnificent palm-groves shading the village of El Risha. Houses are dotted farther among distant palms ... Our steamers are busy out in the waterway; thousands of horses are rollicking on the shore; the tramp of feet murmurs among the palms. We have awakened El Arish from its long, long sleep Morry and I inspected the Turkish fortifications.

February 14th—Information says there are now ten thousand Turks at Shellal alone: and that is only a redoubt.

February 21st—We are riding out to-morrow, fifteen miles to Sheikh Zowaid, then another fifteen to tackle some Turkish outposts around Khan Yunus, and find out what strength of enemy is holding the town. Then hurry back to Sheikh Zowaid. I can see some tired men and horses. But these orders please us. All Bedouin and Arab tribesmen met *en route* have to be rounded up; any running away are to be shot: any approaching our patrols without a white flag or their hands up are to be shot. So the

Bedouin, who fires in our backs, is to be euchred at last. The English general in command must have wakened up. These people are the best spies the Turks have. They have a habit of approaching, in mobs, a solitary man and ripping his stomach open. Now we can fire at them and stop those and similar games.

A family of Bedouins being questioned by a Light Horse Patrol.

32

February 24th—The stunt is over. I wish words were eyes and ears and feelings. We travelled east all day, Cleopatra's old camping-ground awoke to the rumbling of chariot wheels—eighteen-pounders! The desert became harder. Excitement rippled amongst us. We were only a portion of the Desert Column, the New Zealand Brigade and the 2nd Light Horse Brigade, and of course our little mascots—the guns. Presently from the screen away out in front floated a wild shout. Our hands itched to our rifles; we listened for the burst of shell and the scream of shrapnel. Men in the screen stood in their stirrups: they turned and waved, pointing excitely Surely they must see the Turkish cavalry! We rode on, curious as possible. Presently came shouts from the head of the column—men were standing in their stirrups pointing across the desert. Our horses pressed on. Then we saw them. Scarlet poppies! Wild flowers and scarlet poppies. They might have been born of the very love of Antony and Cleopatra: we thought them as wonderful as if they had been. Could this really be the end of the everlasting desert!

On our left was the sea, to our right, the sandhills, Above, the brazen sun. Just before sundown, we sighted picturesque groves of palms, soft and pretty under the setting sun. Stone walls and houses of the east peeped among the trees. The date palms nestled against a golden line of sand-hills by the sea. Fronting the palms was a shallow lake, all rose and silver shadows reflecting palm and sky. Alas, it had the desert sting—it was brackish. Around the lake grew water rushes. How we stared, and exclaimed, and laughed. How eagerly our horses pushed forward to join other columns riding around the lake.

Such was Sheikh Zowaid. It had a proper Sheikh too, in flowing robes and venerable beard, and cut-throat heart. In dignified fashion he strode out to greet the Old Brig. We wondered if he would sell him any hens.

The Turks had installed oil engines and pumping plants. The Arabs were of a different class to any we had seen before.

At 1 a.m. we mounted, and rode silently away. The Desert Column has long since learned to "fold its tent"—which we haven't got—and "steal away" as silently as any Arab. But we vanish to fight. We rode with stripped saddles. No smoking, no talking, just the night with its stars and the soft murmuring of the horses' hooves. Presently came a subdued creaking, and we were riding beside the caterpillar wheels of the Tommy guns.

General Chaytor of the New Zealanders was in charge. We were to surround, if possible, a Palestinian town called Khan Yunus, and capture Sheikh Ali El Hirsch and his band of cut-throats. The Sheikh is a notorious Turkish Intelligence Agent. We've got to find out (you bet we will) if the Turk holds Khan Yunus in strength.

Presently we rode through a village of mud houses, its dim lights all closed in. Then the desert again, and a tribesman's orchard protected by a wall of cactus, its huge, spiky leaves quite prehistoric-looking in the dark. Followed a creaking column of ambulance carts and sledges, all travelling east, east, east! How many times has the Desert Column awakened at dawn riding doggedly into an enormous, blood-red sun! We knew when we passed Rafa by the smell of dead men. The small hours came, numbing our feet and hands. We longed for daylight, its warmth to drive away the cold, its light to dispel the torturing sleep. The horses were travelling splendidly. Hardly a whispered word, but every man was listening, listening, hoping, hoping that the turfy spring, the fresh odour underfoot would not give way to the dead sough of sand. Presently, we heard the rustle of grass! Our horses must have felt it like the kiss of a sweetheart in the dark. In the east a veil of pink deepened to crimson, lightened by needles of steel. We peered at the ground, the ranks grew alive again with the waking day, the longing for a smoke.

Dawn came: we were riding over native crops. The tops of some sandhills nearby were quite pink. Crack! Crack! Crack! Crack! From the sandhill ahead blazed a red rocket, another, another, and another! Crack! Crack! Crack! Crack! Plip-plop, plip-plop, plip-plop, plip-plop. Soon the snipers' rifles merged with infantry in a roll of musketry. The New Zealanders heading the column were dismounting for action.

We galloped over the skyline to protect the right flank. Crack! Crack! Crack! Plip-plop, plip-plop, plip-plop, whizz plop whizz! We thought we were in for it, but it turned out to be merely a clash with outposts. The En Zeds did the scrapping, we merely chased cavalry patrols and guarded the En Zeds from the Turkish infantry entrenched at Shellal.

We gazed towards Khan Yunus. A low ridge sprinkled with trees, miles of cactus hedges surrounding a native town, a fine mosque, and the mediaeval tower of the Crusader's Church built by Richard the First of England. Miles of small ridges under a carpet of grass, the flats and hollows a foot deep in wheat and barley. And poppies, daisies, lilies, irises, cornflowers (I hadn't seen them since a boy), pimpernels, buttercups, and sweet flowers no one knew.

We forgot all about fight until—crack! crack! crack! plip-plop, plip-plop, plip-plop. All the fields were covered with Bedouins, some running at haste towards the villages, others gazing at these strange white horsemen on the skyline, others hurrying flocks of goats, cattle, or camels towards the grim fortifications of Shellal. Dotted among the flats were date groves, and men were there too. Cultivation, flowers everywhere, intense blue sky, a lark singing. We were in the Promised Land.

Our horses had gone mad—the sound of their chewing was one mighty crunch, crunch, crunch. They nosed their master out of the way as he stood before them holding the reins, a man was continually being pushed and nosed and nudged off the grass.

Turkish cavalry patrols came galloping about; fierce looking bucks, all moustache and big leather boots above their knees. Their lance-points gleamed. Our section under Lieutenant Radcliffe formed one patrol against them. Away we went delightedly, the horses reefing for a gallop upon the springy turf, soon putting two miles between us and the main body. An odd bullet whistled past, fired by Bedouins lurking in the barley patches. A horseman appeared from nowhere, leaning over his pony's neck, full gallop for a stone building. We edged away from that building. Another cavalryman galloped to the building. We let him gallop. Silly decoys; they must think us goats! The Jacko cavalry will have to plan shrewder games than that before they lure us on to a machine-gun.

A mile farther on among some trees, were silhouetted seven camel-men. Lieutenant Radcliffe sent Sergeant Solling with Bert and another chap to see what they were, while the remainder of us rode away to the right. Twenty minutes later, Stan and I spied our three chaps coming at a swinging canter with galloping horsemen behind them. As we looked, the leading pursuer leaned over his horse's neck, and puff, puff—revolver smoke! Stan and I burst out laughing—there was old Bert being chased by Turkish cavalrymen, and Bert wasn't wasting time either! The Turks gained, shooting as they came. They yelled as if they wanted our chaps. Like a trick rider the second Turk leaped off his pony, knelt down. Crack! crack! A mighty poor shot! Our chaps didn't increase their pace much, but they had the spurs in.

The Jackos swerved when they saw us and opened out snappily from the shelter of a ridge. And so far as our brigade was concerned it was just patrol fights among those dates and palm-groves, and cultivation, and odd stone houses. We enjoyed it. The smell of green grass and the flowers helped wonderfully.

When the Heads had gained the information they wanted, the two brigades rode back, to the tune of the rear-guard rifles talking to the Turkish horsemen who hung around our flanks.

Passing Rafa, we joked at sight of the boundary pillars, each man calling out when half his horse was in Palestine, half in Egypt.

Some of the Rafa trenches were still half-full of dead Turks. We cursed the Bedouin again for we saw where he had dug up our dead for the sake of the clothes, leaving the poor naked bodies to the jackals.

We arrived late at Sheikh Zowaid, watered and fed the neddies, did the same to ourselves, and then the lucky ones not on duty turned in to dream of flowers and grass all beautiful on solid ground.

February 26th—Reconnaissance to Rafa.

February 28th—Yesterday the brigade rode to El Badari, fifteen miles east. The regiments then branched out—and as a screen scoured ten miles of country. Our troop struck a Bedouin encampment of over a hundred tents, huge sheets patiently made from hair of goats, camels, and sheep. Each great sheet is supported by a few sticks. Whole families camp in them, with goats, sheep, and occasionally a young camel. The shepherds live simply in holes in the ground. These people are not the nomads of the desert. These are "superior." They have flocks of sheep, goats, camels, and a few cattle. They plough the ground a few inches deep. Their plough is a pointed stick. A camel drags it. Their few rude implements are what their forefathers used in the days before Moses. The treacherous dogs fired on us from cover and lit numerous smoke-fires to warn the Turks: they are protected by order of the British Heads. We had patrol fights—one patrol charged us with the lances but they swerved and galloped away when we jumped from our horses and knelt with fixed bayonets.

March 1st—The New Zealanders were shelled at Rafa yesterday, from Turkish fortifications twelve miles away.

March 3rd—Our brigade rode through the winding lanes of Khan Yunus yesterday. Sheikhs in flowing robes and hundreds of Arabs sullenly watched us pass. The old 5th were peppered by snipers when riding by Beni Sela, but A Squadron rode peacefully along the beach to Tel-el-Marakeb. Jacko's cavalry patrols hung around us, popping away at long range. One of our patrols tried to coax them into a trap. Jacko chased the patrol with fixed lances, but he wouldn't chase far enough. The only prisoners we can catch are Turkish infantry. The whole Desert Column is eager to see what this famous Turkish cavalry fights like … We passed over the sites of dead and forgotten cities, finding numbers of coins, so corroded by weather and time that they powdered to pieces when we bent them. There are efficiently built, circular stone wells in this district, some two hundred feet deep. The stone edges are deeply grooved from ropes as the Bedouins haul up the water-buckets by aid of donkeys. The wells must be old as Methuselah, for rope to cut so deeply into rock.

March 5th—A 'plane was shot down before Beersheba yesterday, the airmen fired their machine and ran for it. One of our armoured cars picked them up. Our 'planes report that the Turks have abandoned their redoubt system of Shellal-Weli Sheikh Nurun-Khan Yunus. We are much surprised.

We seem to have the Turks bluffed. We hear, though, that he is hurrying many divisions up to Gaza on the coast and stretching them right across Palestine to Beersheba. Our side is getting a move on for something big too. Our 'planes are fast increasing—we pray they gain ascendancy over the taubes—Australian 'planes are here too. Yeomanry Brigades are coming along from Salonika; infantry divisions are coming from goodness knows where; and artillery is just rolling in. We have even increased the Australian forces.

So far, the Australians and New Zealanders, with our little mascots, have driven the Turk right out of Sinai. The Imperial Mounted Division has not fought as a division yet, though they want to badly enough. They will get it! The infantry have not had a go at all yet. So we are all expecting great things as the battalions come marching in. Now the railway line is across the desert, they can get here. We all see clearly, by our experience in this campaign, how, when war comes to Australia, the mounted men with their wide radius of action, their quick movement and sudden striking power, will be beyond all price against enemy infantry.

...Our 'planes are bombing Jacko day and night ... Some engineers had a narrow escape yesterday, while cleaning out a big Rafa well. A sapper put his pick clean through a plug of gelignite. Examination showed one hundred and forty plugs of dynamite and gelignite with percussion caps, tightly packed between the stones of the well. If the pick had hit the caps—well, the men working below would have needed no burial party.

Sunday—The brigade has been out past Khan Yunus. We combed El Fukhari and Abassan-el-Kebir. Our main body halted at El Belah, but we in the screen went farther. The country has groves of orchards, mostly fig-trees. Thickly populated: big-bearded fellows. Numbers are handsome, scowling brutes; but I don't think they ever wash themselves. They wear belts full of bullets, and revolvers stick out all over them. Under their gowns they all carry at least one dagger, mostly curved. The sheikhs have scimitars as well, attractively ornamented; some have silver handles. They are armed with long barrelled horse-pistols that look as old as a Crusader; but they have many modern German and American automatics too. We found New Zealand rifles in some of their houses.

Twice now we have been surprised on rounding up a mob of these ambush fighters, to find an "Arab" Englishman amongst them. What game chaps these Secret Intelligence agents are. The Arabs torture them to death when they penetrate their disguise. Nelah village looks very queer—the houses have been thatched with earth in which green grass is growing. Their domed mosques, quite white, are picturesque above the earth-baked houses. C Troop rode far ahead, cautious of an ambush. In the fields Arabs ploughed with wooden ploughs, drawn by a camel yoked to a cow. Dotted over the countryside were substantial mud and stone houses enclosed by a loopholed courtyard, with stables inside for the flocks, and hovels for the retainers. Each boss "farmer" is apparently a little feudal lord, sleeping within barred walls, ready at any moment to rise and fight against the desert nomads, Finally our troop halted while seven of us rode away on a wandering patrol.

We approached a two-storied house in a fine orchard, surrounded by a formidable cactus hedge. No one shot at us. Ready for instant flight, we rode up a cactus lane. But no rifle-shot—no blaze of machine-gun. Then a massive door confronted us—smashed. Looked like a desert raid! We dismounted, rifles in hand, and peered in. Silence. We stepped through, tiptoed across the courtyard and came to the house. Its iron-sheathed doors were smashed—its rooms empty—blood-stains on the floor—the broken blade of a cavalry sword. Outside, a huge wooden crane lay splintered. Fallen masonry everywhere, debris, and desolation. A splendid well, too, but one wall of the house had been blown into it. Used cartridges lay about. Listening, we caught the hum of voices. I tiptoed across the courtyard and peeped through a loophole. Women were gossiping under the orchard trees. We left.

33

March 11th—Hamthala. The taubes try less for iron crosses now—though our anti-aircraft guns rarely hit them. We are puzzled at the aggravating methods of the English Heads. Why on earth don't they let either Chetwode or Chauvel run the show. We've driven the Turks out of Sinai and Chauvel has never had a reverse. The English are taking over everything, and in new organizations they put unknown officers over officers who have proved themselves in the field. They are even muddling with our post office.

Abu Shunnar—Good water can be dug right on the beach Reconnaissance out from Belah—sniper duels.

March 14th—A thirty-mile reconnaissance to Weli Sheikh Nuran. The Turks have retired across the Wadi Ghuzze. Providence, not Allah, must have urged the evacuation. From Nuran I gazed at those miles of formidable trenches, each little hill a self-contained fortress approached by gentle rises entirely devoid of cover. The redoubts are immensely strong, the Nuran especially being equipped with deep shell-proof shelters. As horsemen, we examined the outer defences with a queer thrill, for this redoubt was encircled by three rows of "wolf pits." Each pit was six feet deep with a circular rim and a tapering bottom, the lip of each rim adjoining two other lips so that it was impossible for a horse to step between, the pits quite invisible until a man was right on them. We shuddered. If we had had to charge them on horseback, what a thundering crash we would have come, with the redoubt machine-guns mowing us down! As we rode we found that all the larger redoubts were encircled by pits. It would have cost us thousands of men to have taken them. How Jacko must have watched us during our reconnaissance, wishing we would attack! Why on earth did they evacuate? Their redoubts alone would have easily held ten thousand men; goodness knows what numbers the combined trenches would hold.

Hill Nuran itself is a Bedouin cemetery. Jacko wound his tiers of trenches in among the graves, shovelling the occupants out.

Riding back by Rafa, before we knew it our horses were floundering in the loam covering the wide trenches where the huddled dead lay. Curse war.

March 15th—Last night, Turkish raiders swooped on Khan Yunus and drove away the flocks of sheep and cattle. Of course, the Arab sheikhs came howling to us for protection. Good luck to the raiders. Pity they didn't cut all the Bedouin throats.

March 22nd—Something big is doing—busy movement of troops. We ride out in two hours—the cavalry have to break through the Turkish line, then get behind Gaza, and cut the city off from Beersheba. So the mounted men may come between two fires. The British generals intend to storm the town with the infantry. God help the poor chaps who attack the frontal redoubts.

March 23rd—We did not move out after all—fresh troops are continually arriving ... Taubes are active ... The 3rd Brigade is being shelled. Patrol fights on the flank.

March 24th—Fokkers lively to-day. The front has now developed miles long. Guns are boom-oom-oomm-ing. The 3rd Brigade is again under fire. Mounted troops are moving up in columns. Where did England get all her Yeomanry? Fresh Australian brigades are riding by. Long lines of reinforced camelry are lurching over the skyline behind. Battalions are marching, marching, marching—the railway must be pouring them into Palestine. Many infantry have already gone out ahead. What a great hiding the Turk is going to get.

A learned padre tells us that Gaza is a city of forty thousand people. This city of Genesis is one of the oldest in the world. It is Samson's city, the strong man of Israel who when his foes were upon him tore down the city gates and carried them up the hill of Ali Muntar. At Gaza, when chained to the pillars of the Temple of Dagon, he pulled the roof down on top of three thousand Philistines. Delilah was his girl.

Gaza has ever been the fortress city of southern Palestine. Alexander took it, after an historic siege. It became a mighty fortress against Pharaoh's armies, It was a huge city under the Romans. And so on. But what interests me is that the youngest nation in the world will soon be thundering at its gates.

March 25th—We moved out at 2.30 a.m. this morning. When passing Rafa the twinkling campfires lit up the infantry groups hurrying over a frugal breakfast before their long march and—hell! And now came noises partly smothered in the darkness, a continuous squeaking, an oily rattling of caterpillar wheels as columns of artillery were suddenly lumbering beside us. In darkness we rode towards Khan Yunus, an army moving silently, but you could hear it breathing. The murmuring of thousands of hooves, the snort as a horse blew dust from its nostrils, the jangle as a packhorse shook itself, the straining of an ambulance team, the rumble of guns, the whisper of thousands of feet.

A grey dawn broke over Shellal. Morning brought warmth and light. There emerged from the dawn columns of Light Horse, of New Zealand

Rifles, of Yeomanry, of Camel Corps, of artillery, and battalion upon battalion of infantry all with grey faces set towards Gaza. Our thousands off-saddled at Belah oasis. Tense excitement is everywhere—what are we to do next?—this is the 26th. We march on Gaza to-night and attack at dawn.

Late afternoon—An army is advancing—we are excited. We are all hurrying to the little hills to watch them come. *Au revoir*.

It is black night now. The army has come, is still coming, we can hear them marching by. All the Desert Column swarmed to the little crescent hills that line the great plain. Ragged men we are, a little weary, burnt almost black by the fierce suns of Sinai. We gazed into the dust-clouds that floated up between us and sunset. A big cloud, spreading as it lazily rolled nearer was all crimson from the rays. I caught my breath.

We listened amazed, for there floated to us music—Bands! Masses of horsemen took shape. Then infantry! Brigade upon brigade, battalion upon battalion, column upon column, growing rapidly, spreading all over the plain, crushing wide lanes through the barley—the vanguard of the British Army.

As their thousands grew the rhythm thumped up to us—tramp, tramp, tramp, tramp. We gazed down on the closely packed ranks—they looked so well, so fit, so clean, altogether splendid. The rumble of their guns was a hoarse muttering—clink of chains—gleam of wheels—guns, guns, guns Splendid horses, shiny harness, polished chains, rumble, rumble, rumble, and ammunition columns coming, coming, coming!

Parallel were their transport columns, and all the Camel Corps of Egypt seemed lurching red over the skyline.

And away back shone the white-hooded ambulance carts with their red crosses, lines and lines and lines of carts. But the guns, the guns! battery after battery, limber after limber swinging past, swinging past. We gazed down on the guns! What numbers! What calibres! We thrilled as we watched those guns. What things we could do with them! I smiled to the thought of our own precious little guns that had stood by us so gamely throughout the Desert Campaign. But these—why these were GUNS!

Then from all the little hills the Desert Column called down joyful greetings to this splendid army. What boyish faced, cheerful lads they are!

Right across the plain I could hear the laughing shouts in different British dialects: "Are there any Johnny Tourks out this way?" "Yes, and the beggars are waiting for you too!" "They won't have long tu wait chum!" "Good lad, we'll wipe 'em right off the map."

And so on, right until after the sun dimmed down, tramp, tramp, tramp, tramp. It is late now: we feel that Constantinople is ours.

Then a thought already voiced by many of us: "What of the Desert Column? Are we to lose our individuality in this bigger army?" There is a grim regret in the thought, for we are the Desert Column. We move out shortly. I wonder if I will ever write in the old diary again.

March 27th—I wish I could describe even all that I personally saw—but how utterly impossible!

The brigades moved out into a dense fog—we whispered of tiny Dueidar! The ride was a constant stopping and starting with our horse's muzzle rammed up against the horse's rump in front. No talking, no smoking, just darkness and fog and muffled hooves—the smell of horses—the expectancy of a volley—the feeling of unseen life all dense around us everywhere. We wondered where the hell we were; wondered if we might ride into the great fortress of Ali Muntar itself. The horses pressed sturdily on, contentedly, quite realizing the night march but loving the turf under-foot. Occasionally our horses slithered down ravines so precipitous that they could never have faced them in daylight. All a man could do was to lean back in the saddle and hold his breath as his horse's ears disappeared down into the fog. We learnt after, that the horses heading two columns stopped, refusing to budge an inch farther. The riders dismounted, groped forward with their feet and found themselves on the edge of a precipice dropping sheer into the Wadi Ghuzze. Those responsible for the direction of the column did marvellous work, for we passed through the break in the Turkish line between Gaza and Beersheba and pushed straight on through unknown country in miles of fog and darkness.

Came bitter cold. Experience told us when it was after sunrise—but still the fog was a dense white wall. Slowly it began to roll away; we held our breaths; we strained our eyes as men and horses began to loom around. What would face us—No Man's Land or—the Turkish redoubts?

At nine o'clock the fog rolled away. We were riding up a hill; the screen emerged on the crest. Suddenly they disappeared at the gallop and there came a harsh buzzing, distant shouts—two taubes just skimmed the crest of the hill and flew swaying over us, their machine-guns stuttering. They had been camped down the other side of the hill—their pilots rushed their machines as the screen galloped down upon them—the wind as they rose blew off one of the hats of the screen. We captured their mechanics and wagons.

We gazed into the lovely morning. Bare hills all around, smiling valleys and the green Plain of the Philistines before. Away to the left the sombre hills of Gaza, with peeping trees and roofs of the city. A few miles to our front towards the coast were numerous villages set in almond groves, fig plantations, lemon and olive groves and parks of trees.

We hurried to encircle the city so as to prevent the garrison escaping. We could not understand the silence, the fog should have been a godsend to the infantry; we thought they would have walked right into the Turkish redoubts unobserved. As we reached Beit Durdis—crack! crack! plip-plop, plip-plop—but it was only snipers. Some lads galloped out to round them up. Our neddies pushed on willingly. We lit the good old pipes. Crack! crack! crack! ruttut-tut; ssh-sh-ssh—sshsshssh. We halted, and away galloped the supporting troops to help the screen. But it was only a Turkish outpost—it was quickly smashed. We rode on, glancing city-wards at the curious square-built houses peeping among the hills. High above all the roofs loomed a fine minaret. We were calling out to one another that we bet the Turkish Artillery Observation officers would be up there, when boom! a scream, crash! boom! a whine, crash! smoke, red earth, screeching fragments spewed up in front of the leading troop. Where would the next land? There! black smoke and earth right before the column.

Again that shivery little feeling crept from my toe-nails up to the roots of my hair, Blast it! will a man *never* get used to being under fire!

The Old Brig. turned and waved his pipe. The brigade broke into a fast trot, methodically forming into formation to dodge the shells. We wondered and wondered what had happened to the infantry.

I expected we would get particular hell in crossing the big open plain in front. Again came that screeching moan—crash! Morry stood in his stirrups; "My Christ!" he shouted, "it's got the Old Brig!" But as we cantered into the evaporating smoke the Old Brig, cantered out on Plain Bill, still serenely at the head of the brigade.

We cantered by Jabalie. The 6th Light Horse regiment cantered away to block probable Turkish reinforcements coming down the Jaffa-Gaza road. The 7th were in the screen and two troops of the 5th cantered out with them, our section, too. It was an exhilarating canter as we turned in behind Gaza making straight for the broad white road that leads out to Beersheba. Then Australian cavalry were galloping across that plain of the Philistines as we saw trotting along a peaceful cavalcade. Hell for leather we raced, shouting boisterously as the cavalcade whipped up the dust. They were a queer little crowd when we surrounded them, our horses prancing, open mouthed as they tossed their heads rebellious at being reined in. The main captives were in two funny little coaches that might have come out of Queen Anne's reign. There was a body-guard of mostly Arab cavalry all done up like wedding-cakes. But the occupants of the carriages were smart-looking men in natty uniforms. They were a Turkish divisional commander and his staff, moving in to Gaza to take up his command. The staff officers looked like Austrians. I shall never forget the disgusted general, though. His *a la mode Kaiser* moustache was continually a-twitch as vainly he turned his back on the unkempt troopers who threatened him at every angle with all breed of cameras. When he was taken to the Old Brig he complained bitterly of our disgusting coarseness. He demanded some of us at least should be shot. The Old Brig looked as if he was going to explode, but saved himself and roared instead: "Well, it was damned funny, wasn't it?"

Presently, the 7th Regiment connected up with the sea. The brigade and the New Zealanders lined the hills and looked right down into Gaza, its white minarets peeping from the trees that are thick on this side of the town. We halted, because that was the British general's orders. The infantry were to take Gaza, not us. And we watched a tragic day.

The 5th Light Horse camp at Ghoraniyeh, Palestine.

34

The guns were crashing now among those immemorial hills. The Imperial Mounted Division got their chance this day some miles out on our flank, keeping off enemy reinforcements attacking from Huj. The Yeomanry and the Imperial Camel Brigade did their job. The 52nd Division (infantry) guarded us from Turks expected from Khan Yunus way and across the wadi. But we massed Australians and New Zealanders for hours were spectators of the fight. It made our hearts bleed. Here we were gazing right down into the city—and not allowed to enter it. Our position was unique, miles of the semicircle battle was spread like a panorama before us. This was the biggest battle the Desert Column had fought in and yet we watched the main affray more plainly almost than a big outpost fight. We could see the shells bursting over miles of country, see the attacking battalions. With the most bewildered, utterly indignant, with the saddest of feelings we watched the huge bulk of Ali Muntar turning into a roaring volcano, its cactus crest obliterated by the smoke and earth that vividly showed the crimson-black flame of explosions. Toiling across the exposed country at its base, we watched the little toy men of the 53rd Division plodding in waves towards the grim fortress, roaring under its machine-gun fire.

And so, hours late, the first infantry attack developed, and right then C Troop got orders to escort our prisoners away back to divisional headquarters. Away we went, the tired carriage ponies stopping on crossing the ploughed fields, necessitating the officer prisoners getting out and walking, which annoyed them. The general in particular was quite cross. He strode behind the carriage, frowning furiously, twirling his cane, twitching his moustache, snappy to his subordinates.

Stan and I reined in and watched the attack on Ali Muntar through our glasses. The poor Welshmen, coming up the open slopes towards the redoubts were utterly exposed to machine-gun and rifle fire. Shrapnel had merged in a writhing white cloud over the advancing men. They plodded out of a haze of earth and smoke only to disappear into another barrage. It was pitifully sublime. When within close rifle-range line after line lay down and fired while other lines ran past them to lie down and fire in turn. And thus they were slowly but so steadily advancing, under terrific fire. Every yard must have seemed death to them. We could see in between the smoke-wreaths that when each line jumped up, it left big gaps. Some thousands of the poor chaps bled on Ali Muntar that day. And the pity of it was they should have advanced in the fog and been saved that slaughter.

Stan and I cantered to catch up with the distant escort, now trekking among expectant bodies of New Zealanders and Yeomanry awaiting the order to go into action. The rough-looking Australians and En Zeds crowded around the toiling carriages. The nuggety little general was furious: I thought he would twirl his moustache off. He tried to hide back in his carriage but they poked their cameras in the door. He struck at some with his cane; at which one sunburned villain remonstrated: "Aw, be a sport, general!"

The general wasn't—he launched a furious tirade at Chauvel against us chaps and haughtily demanded that Chauvel himself escort him back to army headquarters, as befitted the dignity of a Turkish general. Chauvel smiled. The Turkish general was mad when he had to go back with only an officer as escort.

It was late in the day when we handed the prisoners over. We learned there that Ali Muntar had been taken with the bayonet, but the Turks counter-charged and took it back with bayonet and bomb. Our infantry were trying again. As we rode smartly back we were in company with big bodies of horsemen pouring in from all the hills, closing in on what we confidently thought was the doomed city. Brigades were galloping—the air for miles was all floating clouds of dust and smoke. We watched battery after battery at full gallop from across the open plain, then drawing close to the hills around the town wheel smartly on to their positions, unhook the teams and in a twinkling be firing in salvos. Then across the plain and on up the gentle slopes leading to the redoubts galloped a brigade of Light Horse or New Zealanders, I could not tell which for dust and smoke. Squadron after squadron at a furious gallop with the flash in the puff of smoke above them. More furious and faster grew the flame bursts, the crash! of the shells echoed over plain and hills; there came back riderless horses whinnying as they galloped over the plain.

Squadrons of galloping Yeomanry followed into action—the declining sun flashing on the wheels of the guns—until the plain was a vast cloud of smoke and dust through which could be heard the galloping hooves of thousands, the harsh rumbling of the gun wheels, the faint shout of voices, the neigh of a maddened horse, the crash of bursting shells. It was grand, awe inspiring, but it was terrible! The section stood and listened— we could hardly watch—and the rolling roar of rifle and machine-gun fire rattled from the hills down on to the plain. It was a terrible sight of massed human courage I wonder what other madnesses the human race will go through before the end of the world. We know now that the Ali Muntar was taken and lost three times with the bayonet and then taken again.

The section galloped into the haze to find the troop. We missed them but presently caught a glimpse of the regiment under heavy shellfire, riding out with bayonets all gleaming: we dug the spurs in and leaned over our horses' necks—a New Zealand brigade thundered by as we flew after the tail of the regiment, they were charging into a park of big trees that ran right up into the town. We galloped in among the trees—it was madly exciting. Rifles crackled viciously from cactus hedges, machine-guns snarled from a village on our right. Then we gaped as we galloped straight towards massive walls of cactus hedges ten feet high that ran as lanes right across the park. To our right was only low hedge, and Turkish infantry were enfilading us from there. Lieutenant Waite swerved his troop and the horses jumped the hedge down on to the Turks: we only got a glimpse of that scrap—the lieutenant firing with his revolver, his men from their saddles, until the lieutenant was hit in five places, but what Turks were not killed, ran—while we thundered on and I wondered what calamity might happen when we struck those giant walls of prickly pear. The colonel threw up his hand—we reined up our horses with their noses rearing from the pear—we jumped off—all along the hedge from tiny holes were squirting rifle puffs, in other places the pear was spitting at us as the Turks standing behind simply fired through the juicy leaves. The horse-holders grabbed the horses while each man slashed with his bayonet to cut a hole through those cactus walls. The colonel was firing with his revolver at the juice spots bursting through the leaves—the New Zealanders had galloped by to the left of us, the 7th Light Horse were fighting on our right. Then came the fiercest individual excitement—man after man tore through the cactus to be met by the bayonets of the Turks, six to one. It was just berserk slaughter. A man sprang at the closest Turk and thrust and sprang aside and thrust again and again—some men howled as they rushed, others cursed to the shivery feeling of steel on steel—the grunting breaths, the gritting teeth and the staring eyes of the lunging Turk, the sobbing scream as a bayonet ripped home. The Turkish battalion simply melted away: it was all over in minutes. Men lay horribly bloody and dead; others writhed on the stained grass, while all through the cactus lanes our men were chasing demented Turks. Amateur soldiers we are supposed to be but, by heavens, I saw the finest soldiers of Turkey go down that day, in bayonet fighting in which only shock troops of regular armies are supposed to have any chance. How we thank, now, our own intense training.

The fighting was all in scattered groups; we could only see a few yards around us for pear so I can't tell what happened, only to odd groups. Poor Sergeant Gahn met his death by treachery. He and Lieutenant Scott, Sergeant Hammond and Corporal Ogg rushed an officer with fifteen Turks who threw up their hands, but then seeing how few our men were snatched up their rifles, shot Gahn and bayoneted Ogg. Scott shot five quick and lively; Hammond got one before the rest surrendered.

Lieutenant Graham emptied his revolver and had a lonesome bayonet duel with a huge Turk. Graham got the Turkish bayonet in his stomach. As the Turk lunged to finish the wounded man someone blew his brains out. The Turks only stood it for minutes, they became simply terrified and ran. No wonder! It was they who were being killed, not us. While this was going on, Major Bolingbroke with a few men at Sheikh Redwan, crept up to an observation post that was directing some Turkish guns on our infantry, rushed them, and got away under machine-gun fire with all their range-finding instruments and telephones.

We rounded up the prisoners, sent them back, then carried on backing our way through the hedges and simply killing or capturing any Turks who stood. If they wanted fight, they got it; if they surrendered, well and good.

But A Squadron luckily penetrated through the thinnest edge of the cactus and eventually advanced with the New Zealanders right to the suburbs of the town, while the other two squadrons were still fighting up through the dense heart of the pear. The En Zed squadrons and troops were engaged in the always decisive hand-to-hand fighting too. Before our squadron linked up we heard a hair-raising war-cry and glimpsed the flash of steel among the trees as two En Zed troops plunged into a tiny lagoon towards a scarcely visible trench that was rattling with rifle-fire right until the En Zeds got into them. The Turks met steel with steel but the En Zeds bayoneted thirty-two of them—there were lots of those little trenches hidden all through the peaceful-looking park. The comical thing though was after a squadron of the Wellington Mounted Rifles charged two Austrian guns that, entirely unsuspecting we were so close, were making the park reverberate as they fired over Ali Muntar at our infantry. The En Zeds rushed with an awful yell, the startled gunners and their supports snatched up weapons and fought with frantic terror. Bayonet fighting is indescribable—a man's emotions race at feverish speed and afterwards words are incapable of describing feelings. In seconds only it seemed, forty-two men were stretched dead around their guns: the others kneeled or crouched or lay on the ground stretching up imploring arms. Everything was jolly lively here, we fired down the streets of the suburb, the great mosque was quite close, the bulk of Ali Muntar was reverberating on our left right above us, bullets were ricocheting off the trees, machine-guns stuttering—our fellows were laughing and shouting what they would buy in the city shops.

Turks swarmed to counter-attack and take back the guns—confident big chaps they were. I thought how fine they looked as in massed formation they came roaring out of a street: "Allah! Allah!" "Finish Australia!" with their waving rifles all steel-pointed. Things looked desperate for us little crowd around the guns when away to the right men burst through the prickly pear wearing the grimy felt hats of the 7th Light Horse—one man knelt down in the open, an officer levelled a Hotchkiss over the shoulder of the kneeling man and blazed away, taking the massed Turks in the flank: they fell in writhing masses sprayed by the Hotchkiss bullets, and melted away under the crossfire of the New Zealanders.

Houses were full of lively snipers—one house only seventy yards away was furiously blazing with machine-guns. The En Zeds swung the captured guns around, a corporal swung open the breech-lock and pushed his face there instead of gazing down the barrel, the boys manhandled the wheels and to roars of laughter twisted the gun until the corporal could see a house through the barrel. They shoved in a shell, slammed home the block, fidgeted with the mechanism, until suddenly and unexpectedly bang! crash! up leapt the gun, its wheels spinning, men were flung among the dead gunners but the shell had gone clean through the house and out tumbled twenty-eight Turks with their hands up. We laughed delightedly, the boys grabbed the gun, man-handled it into position, the corporal sighted at another house, the boys held the wheels to keep the damned thing steady—bang! crash! through another house and out ran a lot more snipers coughing from the shell-fumes. The boys warmed to their work, but the third time I don't know what they hit, for I sprang aside as the recoiling gun tried to climb a tree. A major came up bellowing that he didn't want any of his men killed by that "damned Krupp play-toy!"

Turkish battalions were massing in the town, men were swarming down from the fortifications to counter-attack and there was a trench only two hundred yards away blazing into us, but the Turks had the wind up, their shooting was awfully poor. Major Cameron with about fifty A Squadron men under Major Bolingbroke and twenty-five New Zealanders rushed this trench with the bayonet, four machine-guns splaying the trench with lead as we charged. It was blood-boiling work but that charge was quickly over, the Turks would stand up to the steel now only for the first dreadful minute.

As the sun went down, a great shout came rippling faintly, then swelling from man to man right down the New Zealand line: "The Tommies have taken Ali Muntar! Hurrah! Hurrah! Hurrah!" We pressed forward to the town in a wild enthusiasm. Not twenty minutes later there came a staggering surprise—the order to retire! Thousands of Turkish reinforcements were hurrying to the aid of the Gaza garrison and would cut the mounted men off—our infantry were retiring—we were to retire in all haste!

Never will I forget the utter amazement of all troops—we simply stood gazing down the streets of Gaza—officers shrieking for signallers to confirm the order lest it be the work of spies. The sun was right down—repeated signal after signal came: "Retire! Retire! Retire!"

35

The New Zealanders galloped up with spare horses from somewhere and rushed the Krupps away while we all separated and trudged back, seeking our different regiments. How the regiments were scattered! All led horses were some miles back. A quarter moon made the trees and great prickly-pear hedges loom eerily: a weird silence enveloped the battlefield—broken occasionally by shouts of a man calling to his mate among the pear and distant blasts of officers' whistles trying to collect their men together. Once a torch flashed and showed an officer's face all greeny-white from reflection of the pear. But the form he was bending over was a Turk.

We got among a medley of En Zeds and 7th officers who happened to meet. Everyone was mad. The officers swore that if we could not find the horses we should all collect together and march straight back into the town with the bayonet.

It was well after dark when our crowd rejoined the colonel. He was fearfully worried. His squadrons and troops were scattered throughout square miles of prickly pear. The division away out there in the night was concentrating and liable to move off at any moment without us.

We lay down among the pear, awaiting the other squadrons. An ominous rumbling some miles away on our flank told us where the Yeomanry were holding up the Beersheba divisions. We prayed they would hold them up. If all those fresh Turks broke through, then we—

We hoped, too, that the infantry had been able to retreat when we galloped into the back of the town. That was why we did it. The Turks would be compelled to take men from their defences and send them against us, which would give the infantry their chance to get away. But we were so utterly dumbfounded with everything. We knew that at sundown the 53rd Division had actually taken All Muntar and linked up with the New Zealanders—we knew that the troops had really taken Gaza. We just could not understand what the "Big Heads" were doing.

No wonder the colonel, and other scattered colonels, were anxious. Lying there, those of us not asleep just listened, expecting the pattering footsteps of the garrisons in Gaza to come pouring down through the park after us.

That was an awful night. It was our third successive night without sleep too, and our days had been busy and under continuous excitement. While today—

The evening wore on. Suddenly a frightful scream in a vibrating crescendo rang through the cactus hedges.

Sleepers sprang erect, their hair on end. My hands clenched the rifle: it was heart-choking seconds before I could breathe. Hyenas! We knew no man, even in agony, could make such a frightful sound. We sank down again, isolated groups edging closer together. Some men swore softly. A shadowy horse nearby shook a cactus leaf with its trembling. Luckily, our led horses had not been brought up yet. From in among the prickly pear wounded Turks were crying out to us. What could we do! By daylight our parties had collected the accessible wounded and sent them back to safety. But terrified men had crawled right in under the butts of those gigantic pear—and now it was black darkness, the pear stretched in great walls, square miles of walls, all around us.

Suddenly came a fiendish, cackling laugh; another and another—all around us until the air was quivering with the howling laughter of mad dogs. Jackals!

A rifle cracked from somewhere within the pear. We felt jolly glad that that wounded Turk had his rifle.

I don't like to think of those few hours. It must have been after ten o'clock when, at last, the regiment concentrated. How thankfully we emerged from that pear and rode off silently through the trees! Gaza was quiet, but directly behind us that forest of pear was howling with jackals, awful with hyenas.

We crossed the Beersheba road, then turned to go back—to retreat! The fortifications were uncannily quiet but on our left from three directions firing rolled into the night. A few Turkish reinforcements managed to force through some break in the line but one brigade cut and beat them back. The Yeomanry, the 3rd Light H Brigade and the Camel Corps held the Beersheba thousands back easily enough.

We met the shadowy regiments of our brigade at Jabalie, I think it was. I was rolling in the saddle by then. I was aware we had joined up with the division only by the murmur of hooves, the rumble of guns—in my sullen temper they seemed a thousand hearses. We were noisy on this ride: we simply did not, could not care We lit our pipes and matches—miles of men. There was not a shot from the fortifications; there could not have been a Turk in them. We rode right along the foot of the big hills beneath the Turkish redoubts. Not a shot! Out on the left I saw, when I could open my eyes, ribbons of machine-gun flame and stabs of rifle-fire. I hardly heard the reports.

Then everything seemed to dream away into nothing, the steaming smell of horses, the murmur of gun-wheels, the sway of a comrade's body on either side, the pressure of their horses' bellies against my legs, became part of me; all one with the night. The livest parts of us all, something within me insisted, were the bodies of our dead strapped to the limbers. The crash of shells that was really only within my mind vanished; the terrified eyes of a man and the clash of steel, the cackle of a hyena, the howl of jackals, and the crying of wounded men among the pear faded away. The strangest peace filled me as I swayed in the saddle. I could see all our thousands of men and all their dear, patient horses.

When dawn came I was fast asleep. It was bitterly cold. We were still riding. We reached Belah about ten o'clock, having ridden right past the fresh Turkish battalions that were hurrying to the relief of Gaza.

March 28th—This is the date I think. I have very inefficiently told the old diary of what I saw of Gaza. Thank God for the few hours sleep last night. The guns were booming all day yesterday where the Camel Brigade with some infantry and Yeomanry were fighting, but our regiment was not actively engaged. Yesterday evening we filed out to attack again. Thank heavens we slept instead.

March 29th—Guns are booming. I hope it is not going to be Gallipoli all over again. Why on earth don't they mount all the troops and rush in and finish the war? Infantry fighting is too indecisive.

March 31st—Guns booming. Even now that Gaza is over, we know no particulars. We only know that the Turks were astounded when daylight came and they saw that the mounted men had evacuated the rear of the city. Prisoners taken in the fighting since, tell us the Turkish commandant had given his orders to surrender to us in the morning. I suppose it is all what is called "the fortune of war."

We know now that General Chauvel protested vigorously when the Big Heads ordered him to retire the mounted men. We would never have forgiven Chauvel had he taken that bewildering order quietly. The En Zed General Chaytor refused to retire his men until he got the order in writing. Our Old Brig, swore like a trooper; then point-blank refused to move his brigade until every man of us was safely collected.

It is so hard to understand. Our 2nd Brigade and the New Zealand Brigade were actually in the city. The 53rd Infantry Division, after awful casualties, had at last stormed Ali Muntar and the maze of cactus trenches commanding the town and had linked up with the New Zealanders, while the 22nd Yeomanry Brigade were fast advancing.

We just can't understand why our own Heads made us give Gaza back to the Turks.

…We have been wondering at our extremely light casualties (not the infantry's, they lost in thousands—we hear that one battalion alone lost eight hundred out of eleven hundred men) and yet we got into bayonet fighting with the Turk. It simply was that we galloped right into them and in a matter of shortest minutes the individual Turk was fighting for his life in the one phase of fighting that terrifies him—the steel. He went down seemingly half-paralysed with horror at the madness of our rushes. Had we, however, attacked on foot, he would have had the shelter of his trenches and hours to meet us with rifle and machine-gun and shrapnel. In that case, we must have suffered heavily. All the Turks we came in contact with were well uniformed and equipped. They have a new style of bayonet, an awful looking thing, broad steel blades, the backs of them armed with a double row of edged teeth, frightful weapons. German make of course. Jacko tried to use them too, but their biggest men were like schoolboys against us when we got amongst them. The infantry blame the fog for the failure. It delayed their attack for two hours. This puzzles us: we thought the fog would have been a blessing.

...We know very little, except that we have suffered a great reverse. We lost some armoured cars in the fight.

April 1st—Guns booming. And it is such a beautiful morning! We are camped on the shore; the breakers are rolling in. Behind us are green hills, a broad sheet of water near the palms of Belah, groves of fig and other fruit trees. Troops everywhere; some like us, expecting the order to advance at any moment, others coming from, others going to, the firing-line. Quite close is the crackle of rifle-fire, the stutter of machine-guns. Men are dying this morning—writhing on the green grass of the Holy Land! God knows why.

Turkish reinforcements are pouring into Gaza clay and night. And to think that one little Light Horse brigade and one little New Zealand brigade galloped right into the back doors of the town and drove the Turks to their very mosque!

...Neither side is attacking now; just throwing shells at one another preparatory to the next great struggle.

April 2nd—Our anti-aircraft are blazing at the taubes—the air-hawks are persistent this morning. One in particular circled again and again over the great camp quite disdaining the shrapnel-puffs enveloping her. She dived off when a 'plane went straight for her. It is seldom we see our own machines tackle a taube, even now. The taube still seems to have some superiority of speed or fighting strength.

April 3rd—We are experiencing the practical result of yesterday's taube's visit. We are camped on a hill. Below us, spread inland among the hills, are extensive flats on which are the camps of an army. There came a *boom oom-oomm! crash!* and now big-gun shrapnel is bursting above an artillery and an infantry camp. The taubes yesterday spied out the range quite nicely. Now our guns are replying and an artillery duel is roaring over and hill.

…There is wrath in our camp. We have to polish up bits and stirrups. They have seen three years' service and are good for more but some English general thinks we will look more soldierly if our stirrups are bright. The fool does not think of the glint in the sun for a sniper's rifle. So we sit and curse and polish rusty stirrup-irons and gaze down on the flat where shrapnel is tearing the lives from our men, and wonder and wonder whether England really does want to win the war.

…Gaza has developed into a great sit-down battle. We should have shattered it in one day. Ready for the next big attack we have two huge surprises for the Turk, and Gaza must fall. But it should have fallen at a third of the casualties it has already cost. Meanwhile, the guns are speaking, speaking, speaking. And again we are sitting on our rumps polishing stirrup-irons.

36

April 5th—The Egyptian Gazette sports flaring headlines: "Defeat of twenty thousand Turks near Gaza" How can we believe the news of our victories in France, when we read such lies as this!

April 6th—Our section has been scattered all over the troop, Morry, Stan, Bert and I, after living and fighting together for so long. May hell take whoever is responsible. We know the regiment is on the eve of a big scrap too.

…The Tommies tell us they lost forty-two machine-guns at Gaza … Strange that our anti-aircraft guns can never hit the taubes. They blaze at them all day long. On the other hand aircraft are not nearly the weapons they were cracked up to be. They seldom see us if we keep still. It is when we are riding over these bare plains and hills that the taubes see the great clouds of dust. They would be of little account over the Australian bush, but would play hell with the cities.

April 7th—Someone is rubbing the dirt into the section. The sergeant has just detailed me to do a week's guard duty miles away at the dump, because I am "sick," he says. May he grow bald, the cow!

April 8th—There are fourteen of us on guard to prevent the Bedouins robbing the stacks of grain. This morning two hundred of them rushed from all sides. While we were driving off one lot, fifty others appeared from behind the stacks ripping up the sacks and carrying off the grain as fast as they could bag it and run. These are the people the British commander is so anxious to protect, and penalizes us brutal Australians in doing so. These people cut the throats of our wounded; they dig up the dead; they snipe us; they steal everything they can lay hands on. They are the Turks' best spies. German and Turkish officers dressed in filthy Bedouin rags, wander all among our big camps. These grain-dumps are in reserve. But many of the bottom bags have been slit open and their contents stolen. Should we suddenly be thrown back on these dumps there will be no food for the horses, and thousands of men and horses will be lost. But this guard won't let possible starvation threaten our horses not for all the English generals in the world.

This morning is lovely. As far as the eye can see are, in groups, thousands of horses, of mules. Batteries of artillery seem everywhere with tractors and armoured cars and aeroplane camps. There must be fifty thousand or more camels. Close against the shore sandhills are the masts and funnels of the steamers landing our supplies. Close by, is the railway line that has followed us into the Promised Land. Over the vast camp there floats one filmy haze of dust, gone are the bright green fields.

And before us loom the grim redoubts of Gaza. We still wonder why they did not let Chauvel take Gaza that first day, he had the city in the hollow of his hand.

April 9th—This day is monotonous too, it is miserable since the section was split up.

April 11th—The dump has been removed to Railhead—scores of bags were empty, ripped open. The A.S.C. men tell us that when the grain wagons were passing through Khan Yunus the Bedouins used to run alongside and knife the bags, the contents of course spilling out as the wagons rolled along. Eventually the transports had to drive clear of the village. How differently the Germans would treat a similar hostile population! Guns are rumbling out in front; machine-guns rattling.

April 12th—Heavy rifle-fire all through last night. The passing ambulances seem busy.

…The taubes again bombed the aviation camp back at Rafa and are droning up above us now—it is a wonder the falling shell cases from our shrapnel don't kill many more men and horses than they do. They come down with a startling whizz.

…Guns are rumbling like sullen thunder—heavy rifle-fire growls a chorus.

April 13th—There was thunderous bombing at the Gaza trenches last night. This is going to be the Peninsula all over again The taubes have been exceptionally cheeky all day despite our anti-aircraft. Guns are muttering in front but the rifle-fire is just murmuring. Our mounted patrols now are clashing with the Turkish cavalry.

April 14th—I was lying snugly under the old worn saddle-blanket when faintly booooo-oomm! and three seconds later a rapidly increasing wheeze-zee-zee-ze-zezzzzz crash! I hopped out as fragments of high explosive whizzed by. She came again, the deep echoing boom, the crescendo scream of the shell, the tearing crash of the explosion. Jacko had heavy guns trained on Rail Head and he had the range to a pimple. We are surrounded by enormous dumps of food-supplies among which work thousands of the Egyptian Labour Corps. Terrified voices, a patter patter patter—I thought of gigantic rats—as from innumerable lanes among the stacks the Egyptian Labour Corps came running in droves. Boom—oooooo-oomm!—patter patter patter-whee-ze-ze-ze-ze-zezzzzzz crash! Smoke, earth, fragments of niggers, chaos—away across the plain stampeded the niggers followed by a convoy of lumbering camels urged by frantic drivers. These niggers anyway hung on to their camels. We

Australians and New Zealanders and the Tommies on various duties here, stood and watched the Egyptians while some Australian drawled: "Too mean for their land to fight!" The New Zealanders laughed their contempt, and the Tommies understood why Egypt has been under the foreign yoke for centuries.

One big shell ploughed right through a row of hospital tents. That was not Jacko's fault, as a number of our big clearing hospitals are in among the dumps and artillery. The Red Cross should always be away from anyone, it cannot expect sanctuary otherwise. A whistle—and down the line tore the Red Cross train, running away too. She shouldn't have been there.

Other shells came and we feared for the safety of our horses. A solid iron fragment whizzed against the fireplace. The cook cursed real home-like. He ran to the fire and lifted off his pots calling to us to come and get our——breakfast before the——Turks blew it to——hell! So we just waded in and let old Jacko go his hardest.

The big guns have eased off now—our 'planes buzzed off to try and find their battery positions, their anti-aircraft are shelling them now. Some of the poor Tommies were knocked by the long-range shell-fire. The taubes are now laying eggs on the camp—they hurtle down with a nerve-racking whizz and snort.

April 15th—Back with the regiment. To-day, we took the brigade horses towards our grazing-country but the big guns of yesterday opened up and the shells began to burst on the grazing-grounds. We halted. In a nearby hollow were some hundreds of camels with their Egyptian drivers, taking supplies up to the firing-line under Tommy escort. A shell burst among a group of camels. Result: dead camels and several dead niggers; live niggers and camels scampering in all directions. With a thunderous roar another big terror burst amongst a larger mob of camels. Niggers and camels in grotesque horror came racing out of the dust and smoke. Boo-ooo-ooom!—Wheeee-eee-eeeeezz crash! and again a camel-group were obliterated in dust and smoke. Yet again—and we quietly wheeled and rode back, for if those stampeding camels rushed amongst our horses there might be untold disaster. Some of the maddened brutes did bolt into us. However, we are saddled up now ready to move if our particular camp is singled out, for those great shells would play havoc with the horses. At present, we are in a thunderstorm, for our own ironmongery has opened up and the duel is in full swing.

…Last night, I asked our troop lieutenant to let Morry and Bert and Stan and me join up again. He refused point-blank. Because I "don't shave regularly," was the only reason I could get from him. So I went to the major and late that night we were told we could join up again.

...We are marching out to-night—for fight.

About the 18th or 19th—Yesterday was a special gift from hell. About 7.30 p.m. the brigade filed out. We could hear right across the plain the dull rumble of artillery, the murmuring of thousands of hooves, the tramp! tramp ! tramp! of marching columns, shouted commands. We filed past innumerable campfires then out beyond the outposts into the land of the Turk. The night ride wore on as usual, cold and sleepless, made unhappier by a clammy mist. I was shivering when the sun peeped up, thought how warm the damn thing looked glittering above a tree-top when with a re-echoing roar our guns opened the bombardment. All through the morning the big savages roared increasingly until there was thunder rolling for miles. The sun shone bright and peacefully, we watched the spouting earth and smoke on the ridge ahead where was a Turkish redoubt.

Our regiment was in reserve all day. Our job, with other brigades is to prevent the garrisons from Chereiff and Beersheba attacking the forces that are attempting to storm Gaza.

The battle quickly spread for miles. With the warmth, sleepiness vanished, we rode along smoking, winding down among the steep hills that line the banks of the Wadi Ghuzze, a huge old river-bed. Here were congregated brigades of Australians, New Zealanders, Artillery, Yeomanry, Camelry and, by Jove, some of the Engineers with a ton of gun-cotton to blow up some railway bridge. There came that cursed *buzz-zzzz wheeze-zzz crash!* and down hurtled three bombs killing six men and wounding fifteen, and killing thirty horses. If a bomb had dropped on that gun-cotton then a whole brigade would have gone too!

We moved out across a plain towards the hills of Chereiff. The sun grew blazing hot, clouds of stifling dust arose. Again came the taubes; they roared as they swerved viciously down. We dismounted and blazed hatefully while their machine-guns spat down as they roared and rocked above. And with all their noise and hate, they only got one man.

And so on through a stifling day that choked our eyes and nostrils until our throats ran dry with dust. Recently we have been issued with light-coloured Tommy tunics with shiny buttons, they will show up splendidly on the barley-fields for the Turkish sniper. We did not want those tunics and so off they came, though it might have meant our death had a certain happening eventuated. Presently we came upon a huge Bedouin well fair in the centre of the plain. Calling thanks to Allah we rushed it, Aussies, En Zeds, Yeomanry, artillery, and quickly scores and scores of canvas water-buckets were coming up and down that well and the eager horses trod the earth into mud from the spill of the buckets.

The guns were roaring all the dashed day, rifle and machine-gun fire never ceased its rolling backwards and forwards along the hills. We were not in it at all though, our screen rode under shrapnel in the afternoon and soon the wounded were being carried past us *en route* to the dressing-station. Presently we dismounted, expecting our turn. In among the low hills we watched the clouds of dust as bodies of horsemen galloped into action. Some men went to sleep, others lay with their hats over their eyes, watching the shrapnel bursting up among the 'planes. I watched a lark for quite a long time: he fluttered far, far up into the air, singing most joyously.

After dark we rode back to the wadi and the inevitable confusion of great bodies of men watering their horses in strange places.

About nine o'clock, the guns were roaring hell for leather, their flashes all red among the hills. At eleven though, we did not care a damn how they roared for the lucky ones not on duty just lay down and straightway sank into a few hours of the most blessed sleep we have ever had, which is saying a damned lot — in fact, words fail.

A Light Horse Patrol *en route* to Jif-Jaffa, Sinai.

37

This morning, we are still in reserve. The taubes have already laid eggs; we curse the taubes when we shoot the wounded horses. Big guns are roaring away down towards Gaza. The pounding of the lighter batteries is almost drowning the snappish bark of mountain-guns right beside us.

Next day—The brigade mounted hurriedly yesterday and rode towards the Hareira redoubts, where the 6th and 7th opened fire on the garrisons ... Heavy fighting rolled in thunderous waves right away towards Gaza. A damned Fokker dropped a bomb fair on one of our own Tommy guns, blowing the crew and horses to pieces, then he machine-gunned us and we fired back spitefully. And so the long day wore on while we listened to the furious waves of sound as brigades became heavily engaged. At sundown we rode back to Shellal, rationed up, and rode all through the dreary night to El Mendur. No wonder we envy the wounded streaming past. The Turkish defences now stretch right across Palestine from the sea coast at Gaza to Beersheba inland. To-day, our forces are attacking all along the line, before daybreak the air seemed actually trembling from the vibration of the guns in front of us, the 1st Light Horse Brigade has just galloped into action. We are awaiting our turn, watching the Turkish shells bursting upon the advancing troops while our shells are eating up the enemy ridges as far as the eye can see. The blasted Taubes are swooping determinedly—their bombs are exploding with earth-shaking crashes. We are wondering if we have any 'planes left. God help the nation, or rather the civilians, who experience war without having protection against 'planes. They are not much use against fighting arms but they would he hell for the defenceless ... There's a movement among our regimental Heads now—*Au revoir*.

The day after—Yesterday evening was hilariously exciting. At midday the brigade wheeled around to occupy a position from Erk to Dammath. We spread out in squadron troops, picketed the horses, and actually took the saddles off to ease their backs, a thing we rarely do except in camp, then each troop started digging a wee trench, just long enough to hold its own men, thirty. It was in a field of barley, scarlet with poppies. We were too jolly tired and didn't care a damn about digging the trench deep. We were sitting around chewing bully-beef and biscuits when long strings of men appeared on the skyline about two miles away. Those vague horsemen excited our curiosity. We shouted away to the other little troops along our widely scattered line, but they were as puzzled as we. Then appeared mobs of camelmen. We decided they were our own Camel Corps, but were surprised at their peculiar formation. We had just received word that the Turks were retiring all along the line, so never dreamt these chaps could be Jackos. But when three thousand of them started bustling towards us—suddenly the major shouted "Saddle up! Those are Turks!"

What a scatter! Jumping from the trench, throwing on of saddles, horses hurriedly unlinked, gear flying all over the place, then bang! bang!—whee-eeee-eeze-whe-crash! crash! and the horsemen pouring down from the skyline. What a flurry!

The horse-holders galloped our horses away, we grabbed ammunition—each troop rushed for its little trench prepared for the fight of our lives. Our squadron numbered barely a hundred men—a thousand faced us while two thousand faced the remainder of the regiment and the 7th farther along. Other horsemen were trotting over the skyline. Then some mistake was made. Up galloped our led horses: "Retire at once!" we raced for the neddies—bullets whistled by—my heart turned sick when Stan's saddle slipped under his horse's belly—a matter of seconds righted that—we swung into the saddle and were off. "Halt!" and the major yelled that our orders were to hold our position to the last man, so back we galloped and tumbled into the trench. Those clouds of horse and camelmen came straight on—infantry were following behind. We knew then that it was all up, we were hopelessly outnumbered, in the last rush every man would be bayoneted at his post. Thus I made up my mind, and determined to come the hero act and fight to the very last—There was nothing else to do! We opened fire. I got the surprise of my life—those thousands of Turks bolted—cleared for their lives! They reappeared behind cover on the ridges, flaming ribbons shot out of the gathering gloom: bang! bang! bang! bang!—crash! crash! crash! crash! We gazed at one another—a howl of laughter rippled out from troop after troop far down the line—we owned up to our thoughts of a moment before, then blazed away at the silhouetted figures. Our rifles shaking from our laughing.

By Jove, the Turks worked their guns though; the air whistled from the shells. We were thankful for the tiny trench. Jacko must have thought our plain in reality sheltered thousands of men. We felt very lonely, all on our own, and wondered what had become of all the men who had been fighting there in the morning. Where were all our guns that had been shaking the very hills only a few hours before. Then away on our left came the hearty stutter of machine-guns? Welcome sound, our own mechanized terrors blazing away at the Turks. Then the vicious *bang!*

bang! of the Leicester Battery and shrapnel burst above the Turkish guns soon smothering the blaze of their reply.

The Turks would come no closer. We were disgusted at what appeared to us such utter cowardice. Presently came the dark, and orders to retire and form an outpost line. Quietly we rode off, expecting every moment volleys of fire from the darkness. We heard a howl of "Allah" and thunder of hooves away down towards where the 7th would be. Eventually we rejoined the regiment and formed the line against a possible attack at night. Last night I got two hours' sleep. This morning we have started another trench. I am sick of it so I'm just squatting down in the sun talking to the old diary. The boys are too tired to dig, they are sitting up now watching the rolling clouds of dust coming closer from amongst the hills as the Turkish reinforcements march on. The guns are thundering sullenly; 'planes are swooping upon their inefficient squabbles in the sky. We are wondering how the battle is going. We know. Gaza should have fallen on the second day of this our second attack. It hasn't! The infantry were again to take it, the mounted men to hold off Turkish reinforcements from Huj, Hareira, Beersheba, and the Turkish railway line coming right up through Palestine.

We are rather quiet. Sleepless, of course; but we wonder about our grand army. It is very sad—instinctively we know that it has been smashed. There has been bitter fighting all the way to Gaza but we don't know the actual result. We have ceased to wonder what has happened to the two great surprises we had for the Turks. Each must have been a dud else our men would have Gaza days ago … Four taubes have just droned overhead—they are heavily bombing the men behind us—with aerial torpedoes, judging by the terrific explosions. The first six made some of our horses sit back on their haunches. Six of our 'planes have just roared overhead.

Where the dust has been arising from the Turkish reinforcements, is now mixed geysers of black smoke, shattering explosions rake the hills.

…By Jove, the Turks have just brought down one of our 'planes—she dived to earth like a fiery comet—blazing pall for the taube our chaps brought down yesterday … Regiments are galloping out and harassing the Turkish flanks.

…The 3rd and 4th Light Horse Brigades have suffered terrible casualties and the Camel Corps has been smashed up. Two infantry divisions have been almost wiped out, we have heard. I hope it is not true … News is flying. The big French cruiser *Requin* bombarding Gaza, has been torpedoed but she got safely away … Those blasted taubes to-day caused one hundred casualties among the 1st Brigade men and horses.

The day after—At five yesterday afternoon we were relieved. The men started whistling and singing when told we were to have a night's sleep. So we rode through the night until twelve o'clock, rolling in our saddles. However, we rolled off at this place and the few hours we did get has freshened us up wonderfully ... We hear that the En Zeds brought down a taube by rifle-fire yesterday and our own 'planes accounted for another— Hoo-dashed-hurrah! ... The guns are booming, sullen and continuous, but with all our guns the Turks have still more and larger ones. Every gun of the Anzac Division has been in action and by the harsh thunder that rolls right to the sea every gun in the army seems pounding its throat out. By the noise coming from the other side though, all damned Turkey and part of Austria must have their guns in action too. We know now that the Second Battle of Gaza has meant a second hiding for us.

The twenty-five thousand Turks from the sea to Tel el Sharia, and the other big divisions surrounding Beersheba, with the others garrisoning their railway line, must have their tails well up, Our infantry attacked Gaza, Sheikh Abbas, and the Mansura Ridge. The Lowland and Welsh Divisions went into action twenty thousand bayonets strong. We hear there are few of the poor chaps left now. I hope these estimates are not true. The Welsh and Jocks have suffered terrible casualties. The 3rd and 4th Light Horse Brigades with the Cameleers and East Anglian Division tried to break through the Turkish redoubt system between the big Atawineh redoubts and the fortifications on Hareira. They were cut to pieces. The Turks blew the much-boosted tanks sky high.

When the 1st and 3rd Australian Battalions of the Camel Brigade with men of the 161st Infantry Brigade advanced to attack a Turkish redoubt, a tank, The Nutty, waddled along to give a hand. She fought to the very last but got the Australians and British annihilated as well as herself. In minutes only, she was enveloped by bursting shells, the men following her fell in swathes under the terrific machine-gun fire. The infantry suffered frightfully, No. 2 and 3 Companies of the Australians lost one hundred out of two hundred men in a matter of minutes. The redoubt was enclosed by barbed wire, the advancing men could only see the fortifications in patches between drifts in the smoke. They panted on until suddenly the tank began to wobble in circles like an antediluvian monster shot through the stomach. Captain Campbell rushed his six Lewis gunners out ahead and they blazed into the rows of Turkish heads showing above the bullet-whipped parapet. The tank righted herself and under a tornado of shells again clanked straight for the redoubt. The Australians and Tommies fixed bayonets and charged screaming through clouds of smoke. The tank rolled on with her shell-pierced machinery grating and shrieking, fumes sizzling out from the shell-holes in her sides. It was a terrible charge. In the last few yards the struggling monster grew red-hot and belched dense clouds of smoke but whatever was left of its crew either stuck to or had fastened the steering-gear so that the great thing clanked on tearing up rows of barbed wire until in the very centre of the redoubt it burst into flames. Only thirty Australians reached the redoubt and twenty British infantry. They were madmen—the Turks lost their nerve at the blazing tank groaning upon them, at the glint of steel as maniacs burst from the smoke. They fled, six hundred of them! under Germans and Austrians.

Five of the Lewis gunners were dead across their guns. The sixth was wounded and staggered back with his gun, his arm hanging shattered by his side. Six runners were sent back for reinforcements. Four were killed, two wounded. Nine Australian officers reached the redoubt, eight were killed or wounded. Of one hundred and two men in No. 2 Company, only five escaped unwounded. And of all the crowd there appear to be only a dozen who got away. When the Turks counter-attacked, the British officer survivor sent Campbell word that all his men were killed or wounded, he could not get them away, the position was hopeless, he could only surrender. The few Australians who could still run chanced it, and odd ones raced safely through a blazing barrage.

We, but especially the infantry, expected marvellous things of those tanks. They have been screened under palm branches in Belah, awaiting this very attack. Well, the tanks may do great things in France, but they are death-traps to us here.

Our second great enterprise, gas, was a failure too. The atmosphere seems to dissolve the gas immediately the shells explode.

The 3rd and 4th Light Horse Brigades and the Camel Brigade, who charged with the 52nd, 53rd, and 54th Infantry Divisions, say that the Tommies were wonderful—wave after wave were annihilated by machine-gun fire; but other waves came steadily on to be destroyed by the artillery blazing at the tanks, One of the Light Horsemen, badly wounded, was stripped of his clothing and bayoneted to death by the Turks. Most of the Light Horse and New Zealanders lost fifty per cent of their officers. The Yeomanry Brigades galloped almost to the fortifications and fought in a way that has made the crowd accept them as brothers.

…The fight has developed into a great battle raging for miles. We, out on the flank, have got off quite lightly. To-day, big bodies of Turkish cavalry are facing us but they clear out when we dash out after them … What in hell is happening to the infantry at Gaza? The guns are rocking the foundations of the hills there now and we don't know what is happening. Have just watched a Light Horse and New Zealand Brigade galloping into action. The stampeding of their hooves rolled like a thunderous muttering to the roar of the Turkish guns. Horses and men were swallowed in dust and smoke-clouds in which the flash of the exploding shells leaped fire bright … The infantry took the Sheikh Abbas ridges but were beaten back elsewhere. Every available man of the Imperial Mounted Division and the Camel Corps went in on foot to help them … The Canterbury Regiment has been badly bombed … The mounted troops are into it everywhere now, when it is too late, just as in the First Battle of Gaza. Mounted men are fighting from Sheikh Abbas right along Tel-el-Jimmi and in a line of fifteen miles Kh. Erh-el-Munkheibh-Atawineh, at Tel-el-Sharia and Abu Hareira, and along the Gaza-Beersheba road.

We can't understand why they don't let us gallop in as mounted troops and get the thing over. It will have to be done at the finish — after the Turks have fortified all Palestine. Gaza should have fallen on the first day, at a cost to us of only a few hundred lives. Thousands have twice now been thrown away in two attacks and we have been fighting for a month besides. It looks as if we will be fighting for months again after this with the task growing harder and harder.

And all because some English general wanted the honour of taking Gaza by infantry. Infantry cannot fight any faster than they can march. Actually, when fighting, they cover distance practically at a snail's pace. Then both infantries meet, and dig into trenches and advance no more. If both sides had reinforcements and food and munitions, and men had their wives a few miles behind the lines, and each brigade had now and then a fortnight's leave, countries could fight on for ever with no gain to either side.

We hear that the Turkish garrison in the Abu Hareira redoubt numbers six thousand men. If their main redoubts are held in similar strength — and those redoubts have been surveyed under German engineers…?

…Some of our chaps brought in some wounded Staffordshire Yeomanry this morning. They had been lying out all night, putting up a fight against Bedouins who were trying to cut their throats.

38

The regiment rode out last night until one o'clock then did outpost duty against a possible night attack. It is different to the desert outpost. Gone is that silence which envelops all; gone that sense of utter isolation. Away towards Gaza, was the constant muttering of rifle and machine-guns. The hills facing our section outpost were gloomy and silent but to the left the guns belched occasionally, the flash from the big fellows flaming from out the darkness.

After breakfast—The brigade is lolling around, ready for anything … We hear that the navy bombarding Gaza have blown the Mosque up. The Turks were using it as an artillery observation tower. Tommy officers from one of the divisions from France say that in point of numbers engaged the Second Battle of Gaza has been one of the bloodiest in the war. We have lost eighteen thousand men, and gained nothing. I hope their estimate is incorrect. Eighteen thousand is a terrible number of men for a little army like ours to lose. It would be a dreadful loss for a much larger army.

Evening—The sun is sinking. I'm off duty to-night so will write the afternoon patrol. A Squadron alone was reconnoitring towards Shumran. We crossed the wadi Sheikh Nuran and rode across the weird-looking country surrounding it then out on to the great Beersheba Gaza plain. In the distance is a line of big hills with on the Gaza side low hills nearby. We rode towards these low hills, the thirty men of C Troop spreading out ahead as the screen. A large flock of birds arising from the plain right in front of a stone house, drew our attention. Gazing past the birds we saw that the low hills away behind were dotted with mushrooms. We rode inquisitively closer to this large Turkish camp. Decoy bodies of mounted men sought to lure us into a trap, while others again galloped away.

"We'll be getting the whizz-bangs soon," smiled Stan as he made sure his saddle was set for a gallop. Plip-plop, plip-plop; zip, zip; plip-plop, plip-plop, plip-plop; zip. zip, zip—their marksmen can reach us from a surprisingly long range. Their rifles and ammunition are superior to ours. As the screen neared the old ruined house, Stan and I gazed toward it expecting something to happen. A distant bang! whee-ee-eeeezz crash! dust and smoke and plunging horses right where the two little troops rode behind us. Again bang! wee-ee-ee-eezz crash! and another and another and another whining over the screen. That old house was a range indicator for Jacko. Luckily, he used only mountain-guns, the shells exploded harmlessly in the ploughed field, our chaps dug in the spurs and away, the screen wheeling and following them, the shells screeching over us to accurately fall right amongst the lads galloping before. And all the damage done was a few horses knocked and three men wounded. The information we brought back soon had our batteries roaring and now Jacko's hidden camp is being plastered with numerous fat shells.

But here's better news than all the information in the world: we've just been told that, if nothing happens, we are in for a good night's sleep. Bye-bye.

Next day—It was sublimely true. We had a perfect sleep last night. We went stunting this morning, the whole regiment whistling cheerfully. We chased the Turkish cavalry right back amongst their formidable infantry redoubts. They switched their guns on to us from miles around but hit very few. Our fast galloping squadrons completely bamboozled their range-finders. They could not keep us in range for even seconds. Away on the right the 7th Light Horse ran down several of the grey troops and smashed the riders off their horses. We returned in high spirits cantering back four miles under the screeching shells. The horses enjoyed the morning's gallop immensely.

...It is midday now, desert hot. Guns are booming desultorily—Fokkers are humming overhead among shrapnel-puffs. The plain running along the Wadi Ghuzze is under a haze of fine red dust all stirred up by tens of thousands of horses and camels. Far over the Turkish hills are pale red hazes marking where their cavalry are manoeuvring and watering.

The day after—One of the 5th Squadrons cheekily rode within rifle-range of big Abu Hareira this morning. Two Turkish cavalry regiments came galloping with waving sabres, and yelling "Allah! Yah Allah!" chased our boys for several miles. A Squadron is going patrolling this afternoon. All the Light Horse are wishing the Jacko cavalry would stand and put up a scrap against us, just to test us out. They have machine-guns, rifles, lances, revolvers, Arab chargers, swords and centuries of tradition—not to mention Allah—behind them, whereas we have only machine-guns, rifles and bayonets, the good old Aussie neddy, and I suppose the blood of Crusaders, and buccaneers and bushrangers with us. So the odds ought to be on Jacko's side.

The day is steamy hot: water scarce. In the twisted hillocks that form the wadi bank oozes one of God's springs. He, only, knows the generations of men and beasts that have quenched their thirst there. When the Turks held these positions they fitted a water-pipe into the spring, which is convenient for us. Dotted over the country at intervals are wells. It tastes of paradise to us and our horses when after a long, hot and dusty ride we find one of these "tears of God."

…Flies are persistent and in countless billions. The ground is crawling with insects, especially spiders, beetles, scorpions and their progeny and relations. A scorpion bite is poisonous and painful; as we know, for we live either on our horses or on the ground. We have no tents, of course, neither officers nor men. Each man has a waterproof sheet to sleep on, with his greatcoat to cover him. The dashed lice are sharpening their molars. We've never got rid of nor used to them. The approaching summer seems to be getting their back up, or else they've received reinforcements.

…Fighting is continuous all along the dug-in line, day and night. The whole affair is developing into another Gallipoli on a larger scale. Looks as if I'll have plenty of time for the diary: I'll be sitting in a hole listening to the guns for blooming months. The poor old Tommies are very disheartened over the failure of the tanks. They used to say to us: "Wait till th' tanks coom up choom, an' we'll show they bleedin' Turks how to make mince-meat o' they." Well, the tanks came up and "went up." On Tank Redoubt the Turks are using one as an observation post now. They have dug new trenches all around the old iron failure.

May 1st—And so the days go on as continuously as the guns that boom-boom-boom. At Gaza, the blacker the night, the more beautiful the fireworks that burst upon those death-strewn slopes. Upon the grim slopes of Ali Muntar there now lie God alone knows how many more new dead with the sun-dried bodies lying there since that first day of the great fog. Turkish prisoners have told us that they buried two thousand of our dead on Ali Muntar alone after that first day but they got tired and left the others lying there just in the long lines as they fell. Poor beggars. As we have been fighting day and night since, and as our estimated losses for the one main day alone of the Second Battle of Gaza are eighteen thousand men, that army we watched with such pride marching up the Plain of Belah only one short month ago must have dwindled sadly … We hear that our casualties for the Second Battle, as in the First, have been reported much lighter than they really are. What on earth for, we can only guess.

May 2nd—The section were out on Cossack outpost, and scattered for miles to right and left of us at long intervals were other sections, other Cossack posts. And far behind us were the little troops in support, and behind them again, the squadrons, and close by them the regiments, and close by the brigades, then the divisions, and away behind all, the main body and the infantry. And all those many thousands of, we hoped, sleeping men, relying on us little Cossack posts to protect them against surprise in the night. And each post was quite isolated and alone, and very lonely, very alert. And I was lonely, like all the other isolated sentries.

Morry and Bert and Stan lay by me, sleeping huddled up in their great-coats. Our post was not even a shadow in the night, for a tell-tale shadow where a shadow should not be may mean the death of a post. So we had picked our possy between two low rises; I was sitting down cloaked in the natural shadow of both tiny hills. Before me was a clear night view of a plain. Any Turks who might come directly towards our post would sneak up between the rises for they also would never risk being silhouetted over the hill-tops. Thus our post must see or hear them first. Away across the dark plain were low hills all shadowed by big black hills. Among those shadow places and perhaps also out on the plains were the Turkish posts, watching, listening towards us. Perhaps their raiding parties, perhaps battalions, perhaps brigades, perhaps divisions, perhaps an army corps were marching towards us now! Away over those indistinct hills to the right lay Beersheba, old town of a thousand nights. Somewhere fronting us, were the heights of Shereia; dimly to the left the fortress of Abu Hareira, farther still the great Atawineh Redoubts; then, lost in the night, were Abbas Redoubts, Tank Redoubt—goodness knows how many other defences—then grim Ali Muntar and the Heights of Gaza. And around me peace, utter stillness.

But my ears were straining for the rustle of Turk or Bedouin, the muffled tramp of feet, the jingle of a bit, the thump of a hoof! And, by Jove, a man's senses are so trained by this time that he smells the air for camels. And my eyes stared—yet did not stare and did not rest on one spot for long. No jolly fear! And I gripped my rifle, though easily, thumb just resting on the safety-catch, fingers fondling the trigger-guard, the weapon held low across my knees so that the sheen of the bayonet was all dull upon the shadowy ground. Occasionally, very occasionally, the Bedouins do succeed in crawling upon one of our Cossack posts. But what instant wild cats they find our men to be! They have never succeeded in cutting the throats of an Australian post yet.

Guns were rumbling away towards Gaza, the flashes from their red-hot throats lit the hills like lightning swallowed by darkness until the next blazing flash. Faintly came the rattle of rifle-fire jarred by the distinct grunts of machine-guns. Suddenly a flaming rocket shot to the sky. As it paused hesitant, loath to return to earth, there came the crash of bombs, a furious rattle of rifle and machine-gun fire and lightning flamed on the hills. Just a trench raid, maybe a preliminary to some local attack on trench or redoubt, I did not know. I only knew that men were dying in the Palestine night.

Just a Cossack post, I'm blest if I know why I trouble the diary about it. But the morning is beautiful, my old neddy hungrily eats his breakfast, occasionally he nuzzles me with the pathetically empty nosebag, his big brown eyes plainly asking for more. I have none to give him. A lark is singing up in the sky—and here comes the sergeant to detail me for some patrol.

I must scratch myself—the lice are damned bad and I should be lousing my strides like the other chaps, instead of talking to the diary.

Light Horse outpost at El Musallabeh in Jordon Valley.

39

May 4th—It is such a nice morning. Fokkers are droning across a deep blue sky. A lark has shot straight up singing as if its heart would burst in vying with the steely *whirr-rrrr* of the Albatross Scout above. Now the scout is signalling the Fokkers that our 'planes are on the wing. Towards Gaza, guns are muttering where our infantry and the Turkish battalions take a few yards of trench day by day to lose it night by night and take it back next day only to lose it again. And so the war drags on. The Turks have over thirty miles of front now. Both sides are breathing, both are hurrying up reinforcements for another and mightier "push." And the Light Horse cannot but remember that—how long ago?—two little brigades galloped right into the back streets of Gaza.

May 6th—The nights are moonlit in a soft white light. We awake to the racing drone—we lie with tautened nerves while the wings of death roar by bringing *whoo-whoo-whoo-whoo, crash! crash! crash! crash!* and whizzing fragments of aerial torpedoes, stench of fumes, screams of men and horses mingled together. We try to sleep again, and the curse of the night comes again—

Last night, the sky was brilliant with rockets all crimsoned with lightning balls of shrapnel as the Turks sought our 'planes and we sought theirs.

About May 9th—The taubes are playing hell of nights, dropping bombs all among the big camps. Our hospitals should be completely isolated from the fighting camps. Forty bombs were dropped on a big hospital in Belah. Two Tommy doctors were operating on a man for appendicitis; the bombs were crashing all around them, the concussions rattling their instruments; they worked steadily on and finished the operation successfully.

…There have been some changes in our Higher Command, General Chetwode commands the Eastern Force, General Chauvel the Desert Column, and General Chaytor, the New Zealander, the Anzac Mounted Division. We are jolly glad Chauvel is in sole command of us crowd, anyway. We would be more glad still, though, if Chetwode wasn't over-ruled by the Big Command a hundred miles away back on the Canal.

…We have been issued with a Hotchkiss gun per troop. It doubles our strength. A troop of Light Horse are a very deadly little crowd now.

We soon tried the new toy out: crossed the wadi and rode towards Abu Sheria on troop patrol. A mountain-battery banged viciously from brown hills to our right. We spread out and jogged along towards a hummocky ridge from which we could see what was doing, but as two Turkish troops made a dash for that same ridge our troop lieutenant spurred his horse and whistled—we broke into a gallop laughing as the Jackos livened up their ponies. It was an hilarious race, but our horses easily beat the grey ponies. We jumped off, slung our reins to the horse-holders, and ran up the ridge—the Jackos had wheeled in a scurry of dust, galloping for a sheltering ridge. It was snap-shot shooting: *crack-crack-crack-crack-crack-crack!* at the flying target—their lance-points gleaming. As they wheeled around the ridge our Hotchkiss gun opened out. The Jackos' heads lined their ridge and we blazed into one another quite snappily. Just a little private affair of outposts. The Jackos were two to one but our Hotchkiss gun equalized that, though the Jackos in the hills were not sports for they butted in and peppered us with their mountain-guns.

We are deadly shots at half a mile. Our corporal was "spotting" through the glasses he called out and laughed each time we got a man through the head. The Jackos stood it for a while—their bullets flipped up the gravel and ricocheted away screeching. Our Hotchkiss stuttered like a house afire, her bullets ripping lines of red dust right under the Turks' noses. It was a willing go while it lasted—we enjoyed it immensely. I was not exhilarated because I felt gamer, I was just as scared of those bullets as I am in a big battle. Jacko retired at the gallop when he had had enough, leaving seven dead men, all in their big cavalry boots, lying face down the poppies. And it was such a lovely morning!

May 10th—Patrol-fighting is universal for miles. All the mounted men are scrapping; it makes the days quite lively. The 5th do a lot of their patrol work to Hill 410. A patrol may be anything from a section of four men up to a regiment—five hundred. Sometimes we patrol in brigades and then it is called a reconnaissance. Sometimes a squadron, about a hundred and twenty men, patrol out and meet a squadron or a regiment of Turks. There is generally a fight and so far the Turks have lost. Any one of our squadrons will tackle a Turkish regiment any old day, even though the Turks are supported by their guns from the fortifications behind. A Squadron seeks patrol-fights as a butcher's pup seeks meat. B Squadron had a ding-dong set-to a few days ago; chased a Turkish regiment right to Hill 410 under the mouths of the Turkish batteries. C Squadron went galloping up in support and the Turks had the cheek to send out a brigade to chase them away. C Squadron gets most of its more exciting fights from Sausage Ridge and Sihan. All the other Light Horse and En Zeds have their daily patrol-fights and nightly raids and outpost scraps. A 7th Regiment squadron trapped a Turkish squadron in full view of our outpost line a while back. The squadron galloped the Turks down, the Turkish cavalry in a galloping light using their sabres and revolvers but the 7th men just clubbed them from their horses. The Turks are terrified of the very name "Anzac!" And to think of all the preparations we made, all our grim anticipations on meeting these renowned Turkish cavalry!

However, old diary, you have quite enough to do in watching the affairs of a section, let alone the brigade. So it goes on, while the infantry are at one another's throats across the bloody Gaza redoubts. Occasionally we hear the deep-throated navy guns shelling the ancient city while across this great plain we keep the Turk back into his hills, pinning him against his infantry redoubts while his infantry and cavalry have as much hope of driving us back upon our infantry as they have of hopping to Mecca.

...Chaplain Maitland-Woods is a decent old sort. He is quite mad, though; mad on old buried cities, and ancient peoples. Whenever the padre gets a chance, he climbs one of these big old mounds and a crowd congregates, Aussies and En Zeds, Tommies and Cameleers and Artillerists and heavens knows what not, while he holds forth and tells us that the Bedouins were the cut-throat Amalakites who harried David and were just as dirty a crowd thousands of years ago as they are to-day. Then the padre points up this very wadi and tells us of the queer old armies that struggled along it, tough old chaps who tended their flocks and annexed those of their neighbours; who skinned one another alive at times; who built cities that other people razed to the ground. Quaint people who lived and loved and fought and died and vanished within the very dust upon which we lie night after night. So, old diary, if I slug some pages from the past into you, don't be put out, for daily I am breathing these old chaps' dust into me.

The padre has got numbers of the boys archaeologist mad. In their precious spare time they are digging all along the wadi and finding queer old stone houses, and buried tombs and things so musty with centuries that even the padre does not know what they are. The boys disdain anything Roman for that is too modern.

The Wadi Ghuzze is a huge, dry old river-bed cutting right across this Gaza-Beersheba plain. Mighty waters rolled down this dead river in ages past; its precipitous banks are from fifty to seventy feet deep all cut into by sheer ravines and saw-edged washaways. Sombre, weird banks, all grottoed and twisted like a madman's dream. Someone has told us that the man who illustrated Dante's *Inferno* got his inspiration from these sinister banks.

We had terrible difficulty in crossing that wadi. There were only two crossing-places and the Turkish guns from the hills fronting us concentrated on those two crossings. But now, of course, we have crossings everywhere.

Fronting us over the wadi is the No Man's Land of the plain, which we are fast making "our land" against the Turkish cavalry. Beyond the Gaza to Beersheba-line of hills rises, like a great blue wall, the Plateau of Judea, which runs far back behind Beersheba on through invisible Hebron to the great Plain of Armageddon near Mt. Carmel. But that is far away.

From the Plain of the Philistines which we galloped over that day behind Gaza, the hazy blue Plateau of Judea looked very beautiful. It was from there the Israelites gazed down on the land flowing with milk and honey. So the padre says, anyway.

But here comes the sergeant. Bye-bye.

The 8th Australian Light Horse watering their horses in the dust of the Auja Ford.

40

May 12th—Yesterday, the whole brigade crossed the Wadi Ghuzze to deploy as regiments: the screen then rode straight across the plain towards the Turkish hills. Some sixty pounders lumbered behind the main body. Arriving at the first low hills we took possession and formed a line of outposts. C Troop and D Troop were on the farthest hill 630. The ground fronting us sloped in gentle rises and flats to a deep wadi. The country between us and the wadi was dotted with browsing horses and camels, Turks and Bedouins shepherding them, Close by us was a big stone house with graceful trees. Turkish cavalry hid in the house waiting for us to be silly enough to ride up and be shot. On the farther side of the wadi was cultivated country heavily stocked with camels and horses with their attendant Bedouin and Turkish snipers, while riding along the wadi bank, coming and going, ambled the grey ponies of the Turkish patrols. On the low hills behind were their infantry redoubts, with behind them, the Gaza-Beersheba railway and the big Turkish camp of Wadi Sheria. Away to the south-east, we could just catch the sunshine on the white roofs of Beersheba. Behind all, were the big blue hills of Judea.

Our troop leader detailed two sections to form separate outposts closer still to the Turks. The lads cantered out, then galloped to left and right, each post choosing a ridge. Several Turkish patrols accepted the challenge in a whirl of dust, only to lose.

The day wore on. We of C Troop hoped some Turkish squadron would come and tackle us. In the afternoon there was a damned earthquake behind us and an express engine roared overhead. It burst with a crumpling crash in the camp of Wadi Sheria completely obliterating a line of tents under a cloud of heavy black smoke. As more ironmongery roared overhead we laughed at the Turkish infantry running from the camp to their deep trench shelters. Presently the guns knocked off business and were hauled laboriously back. The brigade of course had to wait until they got into safety lest the Turks swarm out and attempt a capture. Then C Troop and D Troop had to wait while the brigade got a move on for home and bully-beef. It was late then. The Turks were working around us, the firing from our small outposts was quite snappy. We ourselves were blazing at the grey ponies as the patrols dashed cunningly closer from ridge to ridge, other outposts away to the right and left were crack, crack, cracking, until at sundown was signalled the distant order "Retire with all Haste!" Our two sections mounted and were away, both troops galloping straight for the setting sun. A light breeze was blowing, the ground was hard, the horses wild for a run; bullets whistled past; it was a glorious ride. We glanced behind at yells of "Allah! Allah! Allah!" "Finish Australia!" We laughed at sight of the little grey ponies coming hell for leather, some riders firing from the saddle, others waving sabres—the rays of the sun leaping from their lance-points. We waved our rifles in laughing defiance; our neddies stretched their necks in an attempt to fly for the vanishing brigade.

Very rarely, in these mad gallops of ours, does any good Australian horse go down. When it does happen, well, the rider lives through the next few minutes anyway.

May 17th—We have moved camp to the Gharbi-Gamli redoubts … Seven New Zealanders were shot on their daily reconnaissance. A long line of Turkish infantrymen craftily moved out in the night and snuggled down in a barley-field. Out ahead of them were hidden snipers. At dawn, horse-patrols out in front were the lure for the trap. The En Zeds chased them into the barley and of course got it in the neck. Yesterday two men in a 7th Regiment screen were killed rather similarly. The Turks are up to all sorts of tricks. But for every man of ours they get, we have averaged up to date ten of theirs. Their taubes have better luck. Today, one dived from a cloud at three of our 'planes that were circling right above us. One of our machines was shot down. English officers tell us that the Fokker and Albatross Scout are superior machines in every way to our own. Well then, it is a shame that the English War Office sends their men up in such obsolete machines. The Papers are full of our air mastery in France, then why don't they send some real machines out here? Surely the lives of flying Englishmen and Australians in Palestine are valuable!

…Our eighty thousand horses, camels, and mules have churned the plains into finest dust, the sea-breeze blows all day and we live in one vast red haze that rises to the skies; it hangs all day long over the earth as a fine red vapour. It is hell when the sea wind changes and a scorcher blows in from the desert.

…We've seen where the papers gave us an odd paragraph or two about the Second Battle of Gaza. Now one has come out with a lot of rot about the "Great Gaza Victory," and glowing lies about the lads out here living in a "Land of Paradise," revelling in oranges, pomegranates, and all the fruits of the Orient. We wish that the idiots responsible for such lies were out here swallowing dust.

May 19th—The En Zeds smashed up a Turkish troop in a patrol scrap a day or two ago, but we hear they were unlucky yesterday losing three men in a little private outpost affair.

...The rations are very lean-gutted lately. Summer has brought the dreaded khamsin. It flames in from the desert as if the world were on fire.

May 23rd—The night before last the Anzac Mounted Division and the Imperial Mounted Division made a raid on the Turkish railway between Asluj and Auja. It was a cheeky enterprise, splendidly carried out. The object was to blow up the railway line as the Turks might any day move their troops along it for a large scale attack against our right flank.

The brigade left camp at 7.30 p.m. the 5th Regiment being advance guard to the column. One division moved towards Beersheba, the other towards Galaze. The all night ride was in choking dust—snort of horses blowing dust from their nostrils—muffled hooves down along the invisible column—thump of rifle-butt against a comrade's leggings.

A khamsin sprang up, it whined in from the desert and its breath was fire—we rode through a haze of hell—I could not see Morry or Stan; I could feel them. The air was saturated with electricity—my little neddy's mane all stiff and bristly. When I touched it, blue sparks leaped through my fingers. A man lost all sense of direction and conscious unity—he felt he was a blind atom in an invisible body moving across the earth.

The regiment started out with a native guide but he soon got lost. The screen went on by compass—when they could see it. At daylight the divisions, spread over some miles, were riding right on to the railway line. The sun glowed through the dust like a molten globe.

Our brigade was disappointed, we were not in the actual blowing up of the line, we just cleared the country of snipers and guarded the flank of the Demolition Troops. The 5th took up a line between Gos Shelili and Hill 1240, rode thankfully on grassy country away from the dust right on to rifle and machine-gun fire. We galloped straight down on the outposts fronting us. One of our lads away out in front had an exciting time. The sun was just bathing the hills—we cheered as he raced over a skyline at the heels of a mounted Arab scout both blazing from their saddles. Neither scored a hit. The last we saw of the Arab he was hitting the dust towards Beersheba. Bang! Bang! Bang! Bang! and our vicious little Tommy guns were surprising the Turks in Beersheba. General Chaytor is a game old beggar. He is very sick but insisted on coming out all that long distance and directing operations from his bed in a sand-cart. About eleven o'clock, our brigade thought we were in for a scrap. Two brigades of Turkish cavalry threatened us but they changed their minds and rode back on Beersheba. Meanwhile the demolition forces, the 1st Light Horse Brigade, the New Zealanders and the Camel Corps, were busy on the railway line. Presently the great Asluj bridge was destroyed, eighteen fine arches of stone, all Ashlar work. The German engineers must have cried to see their masterpiece obliterated in that roaring series of explosions. Then up into the sky went seventeen miles of line with smaller bridges and fine stone culverts, all exceedingly well built. We were all ready to retire by three o'clock. The ride back that night was the very devil. We rode on a big camp where 110 camp had been before. It was a blooming Yeomanry division: I wonder what would have happened had we attacked it! Who would have got the greater surprise? The uncertainty delayed us getting back to camp until after midnight.

…One of our lads got a parcel addressed to "a lonely soldier," Enclosed was a note from the lady expressing the pious wish that a brave soldier in France should get the parcel and not a cold-footed squib in Egypt. The chap who received the parcel sent the lady some photos of our desert graves, with compliments from a cold-footed squib in Egypt.

May 24th—Had an amusing little ride this morning. Quite a common ride: we happened to be engaged in the peaceful task of collecting wood to boil the regimental stew. In this treeless country, we have to rely on the beams in the Bedouin mud-hovels. But the scattered hovels all along the wadi are long since cleaned out, so pioneers must push right to the Turkish lines for the wherewithal to boil our quart-pots. This morning, seven of us went out, hard-boiled old Corporal Nix in charge, with four packhorses. Every hut we visited was timberless. Morry and I rode well ahead as a target to save the packhorses coming under possible rifle-fire. When about four miles out, we noticed a New Zealand outpost on a low hill to our left, and fronting us was a house. Nix and his cautious men edged away towards the outpost. Morry and I rode on, guessing that concealed riflemen must have driven the outpost from the house. But wood is precious, there was always the possibility of some lying round about a house, so we rode on cautiously but hopefully, spreading out and cantering so that probable marksmen would not have too easy a target. Morry yelled: "Blow 'em! if the b—s fire at us well turn around and come home." I grinned.

We drew dangerously close to the house. It was surrounded by a hedge of prickly pear, lovely cover for riflemen. So we agreed on an old, old trick that always seems to work. I suddenly stood in the stirrups gazing earnestly, then yelled and instantly we swerved over our horses' necks and wheeled away at the full gallop. Crack-whizz, crack-whizz, crack-crack-zip-zip-zip.

The ground had been ploughed, our neddies took it at a plunging gallop, knowing the haste necessary as the bullets whizzed by—bullets have a tearing whistle at close range. We cleared the bad ground and took the hard country beyond at an exhilarating gallop, seeing the New

Zealand outpost and our own mates on the little hill and knowing how they must be enjoying the joke. I looked at Morry and laughed, his hard face wreathed in an appreciative grin. I waved my hat derisively at the Turks, confident in the little mare racing away. Suddenly the firing ceased. We imagined we were out of range so pulled up into a canter, laughing, but we had merely ridden into a hollow momentarily out of sight, for as we rode up came whizz, whizz, whizz, whizz, and again into a rollicking gallop. A bullet screeched by my ear, others whizzed between the neddies and stung the ground in front. "What splendid range they keep!" I thought. Then we rode around the little outpost hill and there were the En Zeds and our chaps grinning as they put their glasses away. They had quite enjoyed the fun. They asked if there was any wood in the house? Morry replied; "No, it's in our heads." The En Zeds had an ambulance cart with them. They might have had a job for it. However, we scouted around and had luck. Got just enough wood to load the packhorses.

The guns are booming, machine-guns are stuttering from the Gaza trenches. Damn trench warfare. It gets us nowhere; its monotony is heartbreaking; its loss of life futile.

Turkish dismounted cavalry in action against Australian Light Horse.
The officer (right) in charge is an Austrian.

41

May 25th—Strangely, there is a leakage somewhere. Occasionally we know even a week beforehand when a big move is contemplated The Turks' rifles are superior to ours! in a patrol fight yesterday, as has happened often before, we had to lie on a ridge and take their bullets. It was useless firing back because our rifles would not carry the distance. A spent bullet smacked Byrnes on the back of the hand, embedding gravel into the bruise. He swore like blazes. Their rifle-bolts also are of some type of steel that defies sand and grit whereas our bolts must be continually oiled, they heat very quickly and clog up easily.

…Majors Cameron and Bolingbroke have both got the D.S.O. The regiment is quite pleased.

May 26th—The brigade is to go miles back on the beach for a spell. We are instantly suspicious; we know too well what "spell" means. The truth simply is that all hands have long since broken out in septic sores: a regiment of men rotten with sores makes no longer a hundred per cent efficient fighting force.

May 27th—Scorpions are maniacal fighters. When two are tumbled into a tin they pull one another to pieces. Some of the boys are quite fond of their pets, and carry their prize-fighters with them in a tobacco-tin. They bet on the results of fights. The stakes are mostly wood to boil a quart-pot, tobacco, chocolate (if any parcels have arrived), and money in the case of the affluent. A number of scorpions are "squadron leaders," others even have regimental reputations should they have survived numerous fights. Three in particular are veritable devils. When fit, these demons will be matched to the death. The proud owners are gorging them up with fresh meat and blood; we don't know where they got either from, but there's rumour that one man has been draining Charlie Cox's fowls. "Fighting Charlie" would shoot him. Before the fight, each "trainer" tickles his poisonous pet with a straw. The little beasts clash their pincers in maniacal fury. They are tipped in on top of one another, then should both be torn to pieces, the fight is a draw.

May 28th—The 53rd Infantry Division is marching up to take over our entrenchments. My neddy is ready saddled, gazing with those big brown eyes. Our patrol fighting is with a definite object apart from taking No Man's plain and cowing the Turkish cavalry. We grab advantageous positions away out in front and start the digging of trenches. The infantry eventually take over, while we carry on the same job still farther to the right. So again and again Jacko wakes up to find yet another infantry redoubt facing him.

For if the slow-marching infantry attempted the job they would precipitate a battle and lose thousands of men possibly not gaining these positions by miles. And so it goes on until the grand finale must come and we cannot see the slightest chance of breaking through the great Turkish line except by some terrific mounted charge. And then the mounted troops in their tens of thousands may pour in and the infantry can hurry in after. So we hope it may be, we pray there may be no "Third" Battle of Gaza. But here comes the colonel. "5th Regiment—Mou-nt!"

May 30th—The beach—Tel-el-Marakeb.

June 1st—The authorities are actually going to grant us leave—this way: one man out of every thirty can go to Cairo for five days. One of those days will be spent in getting there, and one in returning. We have figured it out that if the regiment stays here for two years, every man will have had leave. As we are only here for nine days, we miss.

June 2nd—We have been issued with brand new rifles and a more powerful brand of ammunition. We are delighted for now we should meet Jacko nearly on even terms … A 'plane has just been brought down by Turkish gunfire. Both aviators, Australians, were buried at El Arish. A lonely sleeping-place, poor chaps. They have us polishing stirrup-irons again. When will this cursed war cease!

June 3rd—A dozen of us are Ammunition Guard in an orchard in Khan Yunus. Overhead, are taubes circling among shrapnel-puffs. Towards Gaza, big guns are speaking with a steady, rolling roar. Among our orchard trees little birds are twittering. Over the shore sandhills, a sweet breeze is blowing while around us are memories of the Crusaders. Their old castle still towers, sere and grey, above the trees. Everywhere are orchards, enclosed by giant walls of prickly pear. Many are the deep wells of ice-cold water. The Arab inhabitants look dignified in their flowing robes. Some of the girls are desert pretty, and are cheekily inclined towards us but fear their scowling men. The notorious Sheikh Ali El Hirsch and his bonny men have a reputation in these parts, being credited with having cut the throats of sufficient Christians to ensure all the population of Khan Yunus a certain pass to Paradise.

…Tel-el-Marakeb. When off duty, I go for a walk along the beach at sundown. The wet sand is firm, everything is quiet and still, for a precious two hours a man can almost forget the army and claim his soul his own. In a small date-palm oasis Bedouins are camped. At sundown the ragged men walk to the beach, spread their praying-mats of bag, and turning their backs to the setting sun and the sea offer prayers to their God with a faith and earnestness equalled by very few Christians. One evening as I passed behind them a black-bearded chap's eyes lit on me as he stood muttering devotions, but I'm doubtful if at the moment the "Christian dog" existed for him.

I could not help thinking that if they had a religion less dogmatic, less cruel, less intolerant, less murderous, and put their intensity of mind and life more to other things, they might become quite a force in the world instead of being a living dog from the days before Moses.

…On a night raid at Belah, a taube almost skimming the ground dropped only three bombs but they caused eighty casualties to the men of the 53rd Division. Eighty men outed in six seconds! thirty of them in Paradise in fifteen!

Anzac Hostel, Cairo.

June 8th—Land of miracles? Yes! I worked it this way. Paraded to the major for leave, my excuse being that I had an idea of a simple mechanism to turn our rifles into, practically, machine-guns, and wished to lay the idea before the military engineers in Cairo. The major and colonel were keenly interested, and I was told off for special duty in Cairo. So we entrained at 6 pm. yesterday, crowding into a long open truck, filled mostly with Tommies, *en route* to Cairo on a few days precious leave or special duty.

And yet no one would have taken us for men overjoyed. This war has knocked much of the light-heartedness out of us. In the days gone by, that truckload of men would have sung right through the night. Not now! As darkness settled down and we sped back into the desert the men just crouched down on the hard boards. I do not believe there were thirty words spoken throughout the trip. As for me, I could not help it. I lived throughout the Desert Campaign again, the Desert Column, relentless, day after day, night after night, pushing the broken Turkish Army before them across that pitiless desert. How many dreary months in the whole campaign, twelve?, sixteen?—and now I was rattling back over those same sands, in twelve hours!

If ever a man lived life over again I lived over those desert months last night.

How familiar the still, lonely hills were! How bright the stars, how cool the air! What memories those stunted bushes brought up! How merciful the darkness that hid the blazing sands of day!

We rumbled through El Arish, its desert houses all shadowed. What a paradise its palms and water must have been to the Turks struggling back over the terrible sands from El Abd! When we rumbled through Masar I shut my eyes. When last the regiment was there the light-hearted Welsh garrison gave us a rollicking singsong, and now many of the lads lie blackened on the slopes of Ali Muntar. Then on to Bir-el-Abd where the Turkish remnant put up their desperate rear-guard battle, escaping utter annihilation veritably by the will of Allah. What days of alternating hope and despair, of bitter energy those must have been for the Turkish officers! Prisoners told us afterwards that the men lay exhausted on the sands, the spirits knocked utterly out of them by the persistency of our pursuit. The prisoners assured us that had we completely surrounded them during those last days the men would have simply lain on the sand and not moved a rifle-bolt. And we so very, very nearly did it!

Then on past Oghratina with more memories of the stubbornness of the Turkish rear-guard, the deadliness of their concealed snipers. And then, just dimly seen, the broad palm-belts of El Quatia; and I rode again in that great old charge of ours. Then flashed broad daylight of the morning after. There were the Tommy infantry from Romani struggling to keep up with us, collapsing in gasping heaps under the bushes, clawing the brazen sands, going to dreadful expedients to relieve the agonies of thirst. I again felt the pity that spread throughout the column as we gave the poor wretches our water-bottles.

Then returned the nights of fighting with rifle, bomb, and bayonet among the palms and sand-dunes. Such a little army of men holding back a race of desert conquerors, pressing forward, sure of the Canal and victory and the promised lusts of Cairo.

At daylight the train stopped at a very altered Romani. Yes, all through the trip I lived those damned hard times over again and thought what a hard-used poor little beggar I was! I was not consoled by the thought of the millions of men who have perished along that Darb Sultani. I thought glumly of the Roman legions, and visualized the down-hearted Napoleon plodding back there along the sands, swaying forlornly, perhaps, upon a camel. When Christ passed along that track, I don't think He had a camel to ride on.

June 9th—Wicked Cairo. All the morning I've been walking through military offices putting the idea to the right people. To my surprise the task has been simple, and I've been helped quite a lot. I've even had an interview with a major with "Sir" before his name. He has referred me to the rifle-expert so that if the idea passes him, there may be something in it.

June 11th—This rifle idea of mine is going very well. Plans are to be drawn and the idea is to be sent to the Invention Board in London. The English officers here are very decent, and are going to no end of trouble worrying out of the pros and cons of the idea. So something may come of it, after all.

…We hear to-day that the Turks have taken Samson's Ridge. A great many lives have been lost over that place. We hear of a big bombardment of Australia Hill and fighting on Anzac Ridge.

…We have to go back to-morrow. Once again for the tough times, the outposts, the lice, of Palestine.

The 1st Australian Light Horse Brigade riding down from Judea to the capture of Jericho.

42

June 24th—Palestine. The colonel says the best way to prove the rifle is to make a model. I don't know how it's to be done, but what the colonel says, goes. So now, Armourer Sergeant-major King and I are trying to fix up a model in the N.Z. Ordnance workshop in Khan Yunus. I am attached to N.Z. Brigade H.Q. for the time being.

...We have some fast 'planes now, one has already brought down a taube. Events are the same but increasing, as it were! Raids—artillery duels—bombing—rifle fire—cavalry scraps—etc.

A week later—Nine of our 'planes have flown to Jerusalem on a bombing expedition. Queer "angels" to fly over the Holy City. We hear that four arrived back.

...A 9th Light Horseman had stiff luck yesterday. His patrol was reconnoitring. Suddenly, well-concealed Turks opened fire—the patrol turned and galloped but the man's horse was shot. His mate wheeled back but his horse was fear-maddened and the man could not mount in time. He gripped the stirrup and they galloped off but had only gone fifty yards when he was shot through the heart.

...Starvation tucker again. We used to sympathize with the Turks, but the prisoners we capture are all big fat fellows, so we keep our sympathy for ourselves ... The old colonel has been decorated with the *croix-de guerre* by General Baillond of the French Army. The regiment agrees with the honour ... Bomb fighting kicks up a hell of a row in the Gaza trenches at night. An outpost of Scots Fusiliers was completely wiped out by a Turkish raiding-party. A Scotty lieutenant has got the V.C. for bringing in wounded under intense fire. Craig his name is ... The Scotties avenged the raid. A party of them crept out and stormed an easterly post. They only took twelve wounded prisoners: killed the sixty men holding the post.

July 11th—The rifle looks actually like success. Sarn-major King is putting his heart and soul into it. He is a wonderful mechanic. He has only got "skeleton" tools, while some of our material he has cut from a fish-plate on the line—and out of a wrecked taube.

July 14th—We tried the rifle yesterday. The sarn-major was very excited. The rifle fired, ejected the empty shells and reloaded so fast we could not see the bolt work. There were only four separated cases out of nine shots and of course the concussion disturbed our necessarily "tinker" craftsmanship; but decent tools, materials, and a little more thought will remedy all that. I'm blest if the thing isn't going to be a complete success.

…The Turks got two of our 'planes yesterday.

July 16th—There is a wonderful spirit of optimism over the arrival of the new general, Allenby. He has come right out to the line. We hear that he is a cavalry man too. New 'planes, more men, more guns are coming with him. The artillery fire is heavier to-day. Sometimes at night the batteries open up sending rolling volumes of sound crashing among the hills. Raids are nightly. The old regiment is very busy, but as I'm working here on the rifle I can't describe what I was not in. We are within a hundred yards of the mount of Tel-el-Farar. The padre tells us that the Chaldeans also had a fortress on the mound; I wonder if their shades can see our train puffing across the wadi over the bridge the Engineers have built! Clouds of earth and smoke are spouting from the Turkish hills. Allenby's hawks are circling as if they own the sky.

About July 17th—Great excitement this morning. "The Turks are attacking! Every man to the saddle!" So the Anzac Mounted Division swarmed out across the wadi, thousands of mounted men thundering across the plain—guns going hell for leather—neigh of horses—spin of wheels—far-flung shouts of laughing excitement—everyone anticipating a big fight. It was a grand sight. But it was nothing serious. Only five thousand Jacko infantry with twenty guns. They retired to the shelter of their fortresses, and have indulged in an all-day artillery duel with the divisional guns. It is night now. I am back with the En Zeds. The old regiment are at Gos-el-Geleib for the night. "Gos" is the Ziklag of King David.

July 20th—The rifle is almost completed … Guns are booming—taubes wheeling and spinning and fighting—snarling rattle of machine-guns from the trenches—dust clouds across the plain. Morry rode over to see me. The old regiment is out scrapping.

July 23rd—The bombardment around Gaza is furious of nights now, the big hills rumbling as if volcanoes were torturing their bellies. The small hills nearer us are all aquiver under the flashes of the guns. The front has long since developed into a Gallipoli on a much larger scale.

…Last night, a few En Zeds went out on some dare-devil expedition, clad in shorts and running shoes, armed with bombs and sheath-knives and clubs. They haven't returned yet.

July 29th—The En Zeds returned with some pretty lively information. Result, the whole brigade have been out and with the Somerset battery shelled a Turkish camp and played merry hell in general. And so the game goes on.

Heavy rifle and machine-gun fire around Gaza by day and night. By day and night the mounted men harass the Beersheba lines. The Turkish cavalry are quite cowed; we have to chase them right to their infantry redoubts before they will fight. The Turks have brought up a lot more guns. The firing is rolling heavily along the hills for miles to-day.

The rifle has turned out trumps. King and I are to take the model into Cairo to be tested before the Inventions Board ... Morry has just been across for a yarn. He is full of admiration for a Turkish scout. It appears the regiment set a trap for the patrols operating in the El Buggar-Kasif-Karm district. The three squadrons, on a misty night, were concealed as a triangle around a spot where every morning the Turkish patrols ride. Should the patrols ride into the trap, the squadrons were to gallop together and close them in. The only fighting though was done by Sergeant Smith and six dismounted men. The night they rode out these men saw the shadowy forms of Bedouin Irregulars cutting barley for camels. The seven crept up until the Bedouins spotted them and opened fire. Our chaps rushed them with the bayonet and settled the business in seconds. There were fifteen Bedouins. In the morning the trap was nicely laid. B Squadron with a machine-gun troop were lying concealed on a flat hill-top. The Turkish patrols came along and halted, sending forward a lone scout. He could not see a thing but Allah must have whispered in his ear. Instead of approaching the little hill slowly and carefully as is usual he suddenly came out of the half-fog at the full gallop and just flew to the hill-top, glanced once at the men lying there and was away into the mists before a solitary man tumbled to the "joke." Naturally, the Turkish patrols vanished.

August 21st—Cairo. The trial came off yesterday and was a success. The rifle and the plans are to go to London, King and I to Palestine. The O.C. at Zeitoun, Cairo, explained everything. It appears the rifle goes to some experimental machine-gun factory in England. If successful there, a factory will be opened and the rifle manufactured. The O.C. says he doubts that if the rifle is finally accepted, it will be turned out in bulk in this war. King is bitterly disappointed, for at first we had received orders to go to England with the rifle. Last moment orders were for us to return to Palestine tonight. I don't care a damn; I've had a good holiday out of it anyway. Nights of sleep in actually a bed, good food and no lice.

August 23rd—Rafa. Watching a Taube darting at a big fat observation balloon while waiting for horses to take us to brigade camped on the beach. General Allenby has put the whole Anzac Division on B Class. Says they are frightfully overworked. The division is to have a month's spell. Miracles never cease in this land. Meanwhile we are welcomed by the *boom-oom-oomm! boom oom-oomm!*

…Last night an Italian officer from Tripoli was telling us how the Arabs forced three hundred prisoners down a well two hundred feet deep, then filled the well in on top of them.

August 31st—Marakeb beach. No doubt this is a real spell. No idiotic stirrup polishing or such like tommy-rot. Plenty of bathing, sleep, and fresh water. Just enough training to keep us fit. The gas training is interesting. We are weird-looking monstrosities, disappearing into those huge clouds of gas and fumes, to lumber out and then waddle silently into another awe-inspiring cloud. If the Turks give us gas in earnest and we come at them through it I think our very appearance will terrify them. But the gas on both sides has been a failure so far.

…An interesting census of the regiment has been taken. There are just over a hundred original members of the regiment left. Only eight of these (the old colonel is one) have not been away from it. The remainder have all been wounded, some men twice. The bulk of the regiment is reinforcements, half of whom have already been wounded and returned to the regiment. Our tucker could be better. The old regiment put it all over the Turkish cavalry while I was away. I missed some very exciting stunts.

…There is one thing that the ANZACS are all sorry about. The "Desert Column" is now the "Desert Mounted Corps." There has been a huge increase in the army with the new general. We are the Anzac Mounted Division, the Australian Mounted Division, and the Yeomanry Division. Our Anzac guns are the Inverness, Ayrshire, Somerset Batteries R.H.A. The Australians have the Notts Battery R.I-I.A. and A and B Batteries H.A.C. The Yeomanry, the Berks, Hamps, and Leicester Batteries R.H.A. The Camel Corps have been much enlarged too. I don't know what guns are with them. There's a big body of Engineers now and all sorts of other crowds besides. General Chauvel commands the Desert Corps. General Chetwode the infantry corps, and over all, General Allenby. There are other generals too, but I'd be sure to get them mixed if I named them, though the En Zeds would be mad if I did not remember Chaytor. The infantry are the 20th and 21st Army Corps. I don't know all the divisions, but we've met men of the 53rd Division, the 54th Division, the 60th Division, the 75th Division, and others that I don't know. There are Army Troops too, not to mention Flying Camps and Observation Balloons and the Navy. It recalls old memories to us to see the numerous guns of the infantry, especially the whopper fellows, and compare them to our own tiny Tommy guns that worked beside us so doggedly across the desert. We have a warm regard for our own little spitfires.

We know what all these big troops mean. There is a hell of a "push" coming ... Since I've been away, Padre Maitland-Woods has had the time of his life. Some fragments of mosaic were found on top of a hill. With careful digging, the floor of a Christian Chapel was uncovered, done in beautifully patterned mosaic of coloured marbles. A Greek inscription told it had been built in the 622nd year after the Roman foundation of the City of Gaza. The Roman Era began 61 B.C., which gives the date of the chapel as 561 A.D. The chapel stood on the road from Jerusalem to Egypt. In beautiful mosaic is written: "I am the True Vine; Ye are the branches."

The "True Vine" issues from a green amphora of brilliant colours centring a red marble cross, a bright green glory shining from it. A cage with a bird symbolizes the Holy Spirit. On either side is a hare escaping from a hound (the soul escaping from temptation). Around the centre are tigers, lions, flamingos, and peacocks in vivid colours doing homage to the central chalice. It is now called "the Shellal Mosaic."

Well, old diary, I hope I've got the description right. An enthusiastic En Zed gave it me. But there's a joke in it. They dug deeper and came on a tomb containing a skeleton. The enthusiastic padre immediately sent a message to the Base Records Office, Cairo: "Have found the bones of a saint." A wire came back: "Can find no record of this man. Send full name, number, and regiment and identity disk of Trooper Saint."

September 18th—We are moving back to the Front to day. This has been a splendid camp, the only genuine spell we have ever had. Papers arrived lately, stating that five thousand original Australians are to go back to Australia. Of course, the whole division could talk and think of nothing else—until a paragraph came contradicting the first. Whoever is responsible for those paragraphs should be hung and quartered. Better still, they should be made to enlist and after three years' Front Line should read in the papers that they are to go back to Australia for a spell. Next day they should be handed a paper contradicting the report. Of course, the chances are eighty to one that they would not live through the three years. In that case I would be sorry.

43

About October 14th—Fine dust-clouds choking the vast plain: a hell of a camp. Numerous petty duties. Day and night the mumbling of those damned guns—they seem to be breeding since Allenby arrived. Our only happy memory of this camp was the arrival of tinned fruit and milk from Brisbane. I'm sorry the ladies of the Cooee Cafe cannot realize how much we appreciate their gift.

We've just witnessed an air-duel to the death. The spinning birds were in the blue far above Anzac Plain, darting, swooping, tumbling, whirling at one another from above and below, viciously seeking the topmost mastery. We could hear their machine-guns stuttering strangely away up in the air, see the smoke as they blazed away. Tens of thousands of our chaps gazed up at the duel from along the course of the Wadi Ghuzze. Tens of thousands of Turks watched it from the heights of Gaza and Tel Sheria. Suddenly the taube nose-dived, seemed to falter, then with awful speed crashed like a fiery comet. My pity went out to the poor devil in it. That is the second machine the same aviator has brought down within two days. It's high time our 'planes started to score. The taubes have brought down seventeen of ours that we personally know of.

October 22nd—In deadly earnest we are in for it again. We rode throughout last night to Esani (where we must develop water for the infantry) about thirteen miles south of Beersheba. We must find water also for ourselves at Khalasa (Eleusa of the ancient Greeks, holding then sixty thousand people). At Esani and Asluj are great wells, the modern ones completely blown in by the German engineers. And of course, as ever, the one vital necessity for the mounted men is water, water, water!

We know the whole plan of fight; Chetwode's, we believe. The Turkish left flank must be turned by the Desert Mounted Corps and the Anzacs must take Beersheba on the first day. Failure means that the Army will be committed to face a far greater and more terrible Third Battle of Gaza.

Now, old diary. just imagine that this page is Palestine. Past the top of the page is distant Turkey. The left of the page is the Mediterranean. The right of the page is great rocky desert hills. Below the bottom of the page is the Sinai desert. The page is about thirty miles across. We are on the left-hand corner, just across the bottom, facing the line of the Wadi Ghuzze. Facing us on the extreme left edge, is Gaza, with our fleet just down the coast towards us. Right across the page from Gaza, on the right-hand side, is Beersheba at the rock hill desert edge. Right up the right-hand side of the page from Beersheba runs the precipitous Judean hills and the Hebron valley enclosing the long white Hebron road.

So Beersheba is the Turkish left flank which we have to turn and roll their troops back upon Gaza.

It is the boast of the German engineers that the fortresses of Gaza are impregnable, that the redoubts between Gaza and Beersheba are impregnable, and that it is utterly ridiculous even to imagine that mounted troops could manoeuvre to attack, and destroy the infantry redoubts surrounding Beersheba. The hills connecting all their redoubt system along those miles of hills between Gaza and Beersheba are connected by tiers of trenches. The hills from Beersheba running up along the Hebron valley are also fortified. Our Secret Intelligence informs us that recently the Turks have massed numerous guns along the Beersheba front. So the Desert Mounted Corps has a thirsty job to do.

Chauvel is in command of the Beersheba operations. The plan roughly is: the Anzacs must smash Beersheba; immediately word of success comes through, the 21st Infantry Corps assisted by the Navy is to vigorously attack over the sand-dunes between Gaza and the sea, Chetwode is to rush his divisions against the strongholds of Tel Sheria and Hareira, the Yeomanry Mounted Division is to attack with the 20th Infantry Corps, the 7th Mounted Brigade is to attack the southern trenches. Immediately the Anzac and Australian Mounted Divisions break through Beersheba Chauvel is to smash his way right through to behind Gaza.

So there will be merry hell to pay. I wonder if my luck will pull me through!

So we are at Esani, the whole brigade split up into working parties assisting the Australian and New Zealand Field Engineers and the Camel Corps, to dig out the wells. It is marvellous the work we have done already developing system out of chaos. But the success of all the operations depends on water.

October 23rd—Early morning. Rifle-shots, sharp and clear, close by. Mountain-gun fire too. Ours and the Turkish patrols having an argument. Jacko is just waking up that something is doing in the direction away out on this flank. Somewhere north of us is old Beersheba enclosed by its grey Judean hills. There is a warm feeling of expectancy among all of us.

October 24th—Turkish patrols are swarming out from Beersheba. They are fighting stiffly too. Water is pouring into the huge wells we have cleaned out. It was a mighty well-system in the ages when this desolation was a city. The German engineers spent great quantities of gun-cotton blowing in these wells, but we laugh hourly on seeing their work is vain.

October 25th—Yesterday at p.m. we left Esani on a silent and hazardous night march. We got right through, arriving here at Asluj at 4 a.m. The brigade is again developing water; these big wells also are blown in. A taube has just flown inquisitively overhead so I expect things will happen shortly.

…This is a place unknown, giving the impression of quietly sleeping in a world that knows it not. It is built on a tiny flat enclosing both banks of a wadi, hemmed in by sombre limestone hills. It has a snow-white little mosque, and half a dozen Turkish barracks. A commandant's house too with a neatly laid out garden in which nothing grows. The queer place exists by its wells. Huge circular stone wells from which Moses and the Israelites drew water to fill their earthen jars. The Bedouins to-day draw water in the same way having a stone jar attached to one end of a rope, the other to the stern of a patient donkey. The capping stones have cuts worn into them eight inches deep from the ropes of many centuries of Bedouins drawing water. The Turk ran a railway out to this place from Beersheba, the same that we blew up some months ago. He built reservoirs and huge cement troughs preparatory to his triumphal march to Cairo that ended at Romani. With the sun shining on the snow-white hills, this little bit of modernity built on an ancient city, is vividly picturesque.

October 27th—We have been doing some rough patrol work as protection for the well-diggers. The hills are precipitous and rocky, torn by rugged passes in which a rifle-shot echoes like a cannon. The inhabitants are mostly hostile Bedouins. This work is very interesting.

…Our 'planes have won complete ascendancy of the air. All the Army's camps were left standing and given every appearance of being still inhabited for the benefit of any taubes that might break through above. There is a furious bombardment going on towards Gaza. It must be nearly forty miles away but still we can hear the muttering of guns growing into a roar as those nearer Beersheba break out. At night, the flashes are like lightning leaping over the invisible hills.

…The regiment has been on a forty-mile patrol, seeking a regiment of Turkish Lancers reported in the hills near Matrade by Abda, twenty miles south of Asluj. The country was a jumble of rocky gorges. No British had been there before. We got within a mile of Murra and sent our patrols to Abda. But the Lancers had gone. A 'plane was to co-operate with us. She located us, so we thought. The whole regiment in mass, just stood upon a bare, rocky hill gazing up at her circling low. She did not see us at all. Flew back to Asluj and reported they could find no trace of the regiment. So yet again an aeroplane has proved a dud scout.

October 28th—We said farewell to the old colonel, who has been transferred to the 3rd Brigade. I have never seen such a downcast body of men as the regiment who farewelled the colonel by the wells of Asluj. Major Cameron is colonel now, all hands like him. We are lucky in our officers.

October 29th — We've just heard of an outpost scrap away back on our old outpost line. The 8th Light Horse on the 25th seized some ridges facing the big Hareira fortifications. They made a start digging redoubts for infantry occupation. It appears the Middlesex Yeomanry then took over the job, occupying Hill 630, Hill 720, and El Buggar. The Turks came out on the 27th and tackled the posts, sending two thousand infantry against Hill 630 and twelve hundred cavalry with artillery against Hill 720. The Tommies put up a great fight all day and beat back two charges, but in the afternoon Hill 720 was rushed and the garrison annihilated. In late afternoon, the 3rd Light Horse Brigade came up just in time to save the other posts.

…We are on duty day and night here. We do not mind; we realize this water digging as the most critical part of the operations. General Allenby visited us. We thought quite a lot of him coming out all this distance and seeing with his own eyes what is being done.

…The fight is spreading along the whole line now. There is a heavy bombardment of Gaza. Everything possible is being done to attract the Turkish attention from us. They believe the main attack will be at Gaza. Meanwhile we crowd at Esani and Asluj are working night and day to supply a reservoir for tens of thousands of men and horses. When that is done, the mounted troops will mass at the water, there will be a night march, then — ?

November 2nd — Here goes for the great fight and the grandest charge of mounted men in history. English cavalry officers are now swearing it was so, anyway.

Loaded with three days' rations, our horses' bellies full of food and water, we left Asluj on the night of the 30th. A full moon shone silver-white on a metalled road running between the hills. How different to the desert night marches was the hoarse rumble of artillery, transport, and ambulances along that hard road! We knew it led to Beersheba. All were excited, each in his own quiet way. I had that queer, almost subconscious fear I have when I know that soon I will be under heavy fire and that we must advance into it. We were the Anzac Mounted Division, marching twenty-five miles from Asluj to the assault on Beersheba. The Australian Mounted Division we knew were somewhere among the hills marching thirty miles from Khalasa to Beersheba to add their weight to the assault. The 7th Mounted Brigade were to march direct from Esani and mask the southern trench system of Beersheba. The 60th and 74th Infantry Divisions were to attack and capture the outer defences on the west and south-west of the town. As for the operations extending right away to Gaza—well, everything was on such a vast scale that we simply knew the Desert Men had to take Beersheba on the very first day, else the whole battle was lost.

So we rode along, very alert when the road branched at midnight. Up one road went the 7th Light Horse, to tackle a Turkish outpost astride the road at Bir Ara. A New Zealand regiment rode up the other on similar duty. The main body followed, the valley echoing to our thousands of hooves. Just at daylight came the sharp crack of rifle-shots, the stutter of a machine-gun. The horses pricked up their ears; we knew the outposts were being smashed. Glad we were when the sun flooded the hills. Away went the longing for sleep, out came pipes and cigarettes along the column. Mates looked at one another with a half-smile, musingly. The valley grew wider and wider, Bedouin cultivation made its appearance in ever-increasing plots. Then, five miles away between the hills we caught sight of the white mosque and houses of Beersheba.

Brigadier General Granville de Laune Ryrie leads the 2nd Australian Light Horse Brigade across the desert at Esdud on the Philistine Plain.

44

The attack had already started. Above the far-flung redoubts floated shrapnel-puffs while clouds of smoke masked the trenches. The shells exploded sharp and clear—machine-guns stuttered with a steady, businesslike precision—rifle-fire cracked briskly.

We had a hasty breakfast at the well of Hanim where the order came: "Move with all speed and seize Tel Es Sakaty." As the brigade cantered out of the abrupt Judean hills the battlefield unfolded like a panorama—a four-mile wide plain, then the low hills fronting Beersheba, and running away to our right the white Beersheba-Hebron road between frowning hills. We deployed in artillery formation to dodge the falling shells. Hiding in a depression behind the hills was Beersheba, the white dome and minaret of the great mosque and the railway station, barracks, and numerous buildings growing plainer to us. Clouds of dust swirled around the town as motor lorries loaded with infantry raced away up to the hills to reinforce the fortifications. Bodies of cavalry were galloping to the threatened points. We broke into a fast canter, our little Tommy guns bounding down and up the gullies. Then, as far as the eye could see were our own troops pouring from the hills on to the plain until they were moving regiment after regiment, brigade after brigade, in dust-cloud after dust-cloud—all moving steadily towards the Turkish hills. The plain was hard, there grew a muttering that spread for miles—the pounding of ten thousand hooves with thunder in it from the wheels of a hundred guns.

It was a grand sight, the thrill, the comradeship, the knowledge that soon hell would open out, filled us all, I know, with the terrible intoxication of war when the movement is rapid. I was scared for I understood what was coming, though most of us laughed when the first shells screamed towards us, other men smoked as we broke into a thundering canter holding back in the saddles to prevent the horses breaking into a mad gallop. I think all men get scared at times like these; but there comes a sort of laughing courage from deep within the heart of each, or from some source he never knew existed; and when he feels like that he will gallop into the most blinding death with an utterly unexplainable, don't care, shrieking laugh upon his lips.

From then on our time was fully occupied by events around the brigade alone. Soon we would be too occupied to note anything beyond the regiment, then it would be the squadron, then the troop, then very likely the section.

The 7th spread away in a racing canter—they swept among the mud houses of a Bedouin village to the rush of cowled figures, scattering camels, donkeys and goats. Shrapnel screamed over them and burst with a splitting crash, but as ever the Austrian gunners could not keep pace with the target. As a side glimpse, we saw Bedouin camels, goats, and squawking fowls falling under the shells meant for the 7th. Shrapnel screamed behind us, a high explosive crashed earthwards with shattering force. We broke into a hand gallop with the canary whistle overhead as the first machine-guns sought our range. Then the 7th galloped down a convoy and passed straight on. It was a glorious ride. Some of our led horses became so wildly excited that improperly strapped on packs rolled under their bellies, but the regiment galloped on, and I did not envy the unfortunate horse-leaders who had to dismount with their frantic horses and resaddle those packs all alone on that shell-thrashed plain. Suddenly a Turkish cavalry regiment with four mountain-guns appeared almost in front of the 7th, evidently coming out to seize the Hebron road. They stared amazed, then wheeled and galloped back hell for leather. They had too much of a lead on. They got in behind the Tel Es Sakaty redoubt, thrashed their gun-teams up a hill and wheeled around in action just as the 7th and we plunged down into an old wadi-bed facing the redoubt.

We jumped off, handed over our horses, climbed the bank and then the brigade was well and truly into it: each man had all he could attend to in trying to shoot Turks and not be shot himself. Rifle and machine-gun fire grew into a steady roar, the air was one continuous whistling hissing as if thick with vicious serpents; the ground spurted dust and flying pebbles and splintered bullets. The Ayrshires galloped their guns into action with a screech of wheels and tattoo of hooves firing from the very dust into the cavalry regiments' guns and simply blowing them out of action.

Soon all the battlefield was under clouds of fine red dust. The roar of the guns rolled over the low hills and most peculiarly, I could hear the re-echoes crashing among the rocky passes away towards Hebron. There was not a breath of wind. The Turkish redoubts seemed floating under dust and smoke in which the flash of exploding shells was dancing. Through all the inferno came the reverberating roar of our "heavies" engaging the Austrian howitzers.

The morning rolled on bringing its heat, its hot rifle bolts, its thirst; longingly we thought of the cool wells of Beersheba, and by Jove I know I experienced a choking feeling of the senses on remembering that we *must* take those walls. Now and again I indulged in a bo-peep at this fascinating scrap. Once, from our little rise, I saw lines of horsemen galloping out of red mist away to our left. They were the 1st Light Horse Brigade, the Inverness Battery furiously covering their gallop across the plain. The Somerset Battery came galloping with the sun shining upon their spinning wheels. They wheeled in splendid order and almost instantly the guns were flaming through the clouds of dust actually within close rifle-range of the Turkish redoubts. About midday, our brigade took Tel Es Sakaty. The Turkish survivors were exhausted, their faces streaming with dust and sweat and blood. Immediately, orders came to brigade to seize the Hebron road as 'planes had reported motor-lorries of reinforcements streaming down from Hebron.

As the regiment mounted, several of us were hurriedly detailed to remain on an observation duty. The brigade galloped off and soon were astride the road. We, on observation, climbed a hill and watched the battle for the remainder of the day. It was all hazily distinct so far as the eye could visualize though obliterated again and again by rolling clouds of dust. Away to the left the New Zealand Mounted Rifles were having a hard fight to take the Tel el Saba redoubts. The machine-gun fire just roared from down there, our artillery all along the line were thundering at the German machine-gun nests. As the afternoon wore on we watched the 1st Light Horse Brigade fighting their way around the flank of a redoubt. Taubes were roaring all over the fortifications, the plain, the wadi and the ridges, their heavy bombs exploding in series of smashing roars. Through the glasses, we watched them bombing Chauvel's and Chaytor's headquarters four miles away where the generals directed the battle. I wondered what their thoughts were for all the operations, apart from the dust, were spread plain before them. Chauvel must have been terribly anxious as time wore on for if we did not take Beersheba by nightfall then we must retire to water thirty miles away and the infantry divisions now in action right to Gaza would be in a terrible fix.

We saw that grim work would soon be doing on Tel el Saba as the 3rd Brigade came galloping up to reinforce the En Zeds. We watched excitedly as we saw the New Zealanders, like little men, advancing in short rushes. Then farther along, the 1st Light Horse Brigade began advancing in bent-backed rushes. Machine-gun, rifle, and artillery fire increased in fury. Then we caught the gleam of bayonets—we strained our eyes as one line of men were almost at a trench, they were into it—faintly we heard shouts as line after line surged on. Quickly the firing from Tel el Saba itself died down. Then we saw it was taken! We just laughed—we were jolly glad. Time rolled on. The outer defences were ours but Beersheba still held out. It was almost sundown, and by Jove we wondered what was going to happen next. The 9th Regiment and its machine-gun squadron were heavily bombed, the New Zealanders got hell from the taubes, while others flying low spread death among the 8th Light Horse. We heard afterwards that their V.C. colonel was among the killed.

Then someone shouted, pointing through the sunset towards invisible headquarters. There, at the steady trot, was regiment after regiment, squadron after squadron coming, coming, coming! It was just half-light, they were distinct yet indistinct. The Turkish guns blazed at those hazy horsemen but they came steadily on. At two miles distant they emerged from clouds of dust, squadrons of men and horses taking shape. All the Turkish guns around Beersheba must have been directed at the menace then. Captured Turkish and German officers have told us that even then they never dreamed that mounted troops would be madmen enough to attempt rushing infantry redoubts protected by machine-guns and artillery. At a mile distant their thousand hooves were stuttering thunder, coming at a rate that frightened a man—they were an awe-inspiring sight, galloping through the red haze—knee to knee and horse to horse—the dying sun glinting on bayonet-points. Machine-guns and rifle-fire just roared but the 4th Brigade galloped on. We heard shouts among the thundering hooves, saw balls of flame amongst those hooves—horse after horse crashed, but the massed squadrons thundered on. We laughed in delight when the shells began bursting behind them telling that the gunners could not keep their range, then suddenly the men ceased to fall and we knew instinctively that the Turkish infantry, wild with excitement and fear, had forgotten to lower their rifle-sights and the bullets were flying overhead. The Turks did the same to us at El Quatia. The last half-mile was a berserk gallop with the squadrons in magnificent line, a heart-throbbing sight as they plunged up the slope, the horses leaping the redoubt trenches—my glasses showed me the Turkish bayonets thrusting up for the bellies of the horses—one regiment flung themselves from the saddle—we heard the mad shouts as the men jumped down into the trenches, a following regiment thundered over another redoubt, and to a triumphant roar of voices and hooves was galloping down the half mile slope right into the town. Then came a whirlwind of movements from all over the field, galloping batteries—dense dust from mounting regiments—a rush as troops poured for the opening in the gathering dark—mad, mad excitement—terrific explosions from down in the town.

Beersheba had fallen.

45

Next morning, our mobile regiment was detached for reconnaissance duty up the Hebron road down which old man Abraham had travelled to Beersheba. Although without sleep we rode cheerfully into the Judean Hills, for the Desert Corps had made good. Behind us and far to the left roared the fight along the line of Ras el Nagb-Tel Khuweilfe redoubts where the mounted troops were hurrying to help the 20th Infantry Corps. Even the air we breathed seemed impregnated with the fever of great events. We had no idea of what was happening away towards Gaza: we guessed. The sky was blue as blue could be, but tense, throbbing under a fierce, cannon thunder-storm. But we of the Desert Corps had turned the Turkish left flank and captured the precious water. Now we could deal him hammer blows and break his fortifications, redoubt after redoubt. He must eventually crumble and be rolled back upon Gaza. So the regiment was in a rollicking humour. The hooves rang on the metalled road, wildly picturesque in the towering Judean hills. Then a five-mile ride, bang! whe-eee-eezz, crash! and earth spouted up just ahead of the leading troop who scattered like startled hares. The air was so clear that I noted the blue wisps from their pipes being overwhelmed by the dust and smoke. The regiment cantered under cover, musing on what strength these Turks were. They held a hill-top, seemingly equipped with only two mountain-guns. Their riflemen aggressively opened fire, our advanced troops answered briskly. The rifles rang out sharp and clear in the rocky valley— bullets struck boulders smack! smack! smack! smack! but no machine-gun spoke. For some hours we watched them and they shelled us when, rarely, we gave them a target—we are long since "war-wise." So long as no heavy reinforcements came marching down from Hebron and Dharayieh to strike our fellows fighting behind Beersheba, everything was all right with the regiment. From a craggy peak that overlooked that weird country, we on outpost saw distantly, screeching like a desperately wounded bird, a taube spin down and in frantic efforts to right its drooping wings crash among the hills. How times have changed since Abraham was a boy! The Taube had swooped low to bomb the 8th Light Horse but the lads' volleyed bullets had avenged their colonel.

Presently four of our armoured cars came spinning merrily along that old road. They chugged within nice range of Jacko, but he kept quiet as a mouse. We had just time to gallop a man down and warn the motorists that they were riding straight into a Turkish ambush. They wheeled around and we laughed. It seemed seconds only before they were dust-clouds flying towards home and Beersheba. Jacko must have cursed us interfering spoil-sports.

Delightedly we found pools of water in an old wadi bed. Heaven smiled on the Desert Corps when she sent that thunder-storm some four days ago. We boiled our quarts and got right into the bully-beef and biscuits—and just didn't they taste well!

Then the regiment craftily reconnoitred that apparently easy capture ahead—and discovered that the steep hillsides all along the Turks' side of the valley were lined with well-concealed trenches, heavily manned. They opened out on us and that old valley echoed to splitting volleys. But the regiment had abundance of cover. We fired back from behind the rocks, joking at the discomfiture of the Turk. These little interludes are a welcome chance to write up the diary ... At nightfall we rode back to Beersheba and blessed sleep.

The whole brigade rode up the Hebron valley this morning. Soon the Turks held up the 7th Regiment so we turned in among the hills, rode up gullies and, climbing, managed to keep a range of hills between us and Jacko.

Evening—We are dismounted, cautiously lining the hill-tops, looking away down at the Hebron road below The Turks are on the black hills which tower like a wall along the opposite side of the road.

November 3rd—Two of our Ayrshire guns have been indulging in an all-day duel with an enemy battery of 77's. One of the Ayrshire's has just received a shell through her limber. Distinctly comes a tremendous bombardment over towards the Tel el Rhuweilfe fortifications. There is stubborn fighting at Hareira and Sheria. We don't know what is happening towards Gaza, the muttering thunder from there is ceaseless day and night.

November 4th. Evening—very interesting today. The regiment moved out, riding parallel with the Turkish positions. The sombre hills are all gorges and frowning bastions of rock fissured by tortuous ravines. A stunted green-grey bush covers the rocky slopes. It is ideal country for ambuscades. On top of bare hills, holes have been cut down into the solid rock, some are fifteen feet deep, all are chambered underneath. The sparse rain runs over the bare rock and collects in these cisterns. The Bedouins cover the little entrance hole with a slab of rock so artfully that one would never dream there was water below. The cisterns are nearly full now due to the recent thunder-storm, That storm has proved more help to the Desert Corps than another division of men would have been.

Our screen and flank-guard ride tensely alert, for Turkish snipers remain concealed until the screen ride almost to their rifle-muzzles. Four of the 6th Regiment on screen duty just now rode up to within thirty yards of snipers—one man got away.

Next day—We are not riding out today, but the Turk has spoilt our much-needed rest. His shells chased us early this morning from the wee piece of flat ground we were camped on. Now we are away up among the rocks, our sleepy eyes watching his shells dropping uncannily close behind us. We are quite invisible to the Turks but of course we understand that the "friendly" Bedouins are signalling the Austrian gunners.

Evening—We watched a fierce little artillery duel this afternoon. The Ayrshire guns unlimbered behind our hill—their grim muzzles looked skyward—they belched and their shells screeched overhead searching for the Austrian guns. They found them and the challenge has set these immemorial hills reverberating to the crash of shells and the spiteful bang! bang! in reply. For a while, the Turks searched, then gradually their shells crept nearer and nearer our guns until they exploded a hundred yards behind the battery, then a hundred yards in front, then slowly, uncannily, they began to close in on the distance, creeping nearer and nearer. The gunners kept steadily firing their little spitfires. Then earth and smoke burst up before the gun-muzzles enveloping the weapons, but out of the smoke spat their flashes in reply. Then came a hail of shrapnel that lashed the gun-shields while high explosive burst in fragments that screeched against the wheels and shields. The gunner survivors worked their guns without the slightest wavering but presently came "Cease fire!" The gunners are now crouched under the guns which are being hailed by splintered rocks, flying metal and shrapnel: darkness may save them. We are expecting momentarily a high explosive to make a direct hit and send guns and men to smithereens.

The day after—Morry, Bert, Stan and I spent last night crouched in a cave across on the Hebron hills. It was a tense little night with the brilliant stars and cold rock slopes for company—and the Turks who knew not that we were there. But the Turkish patrol we wished to ambush did not come with the dawn. At daylight we crossed the valley and hurriedly climbed our own whopper hill. The rising sun lit up the crowns of the Turkish hills, the peeps at Hebron's white houses being quite pretty ... Our battery changed its position and again it and the Austrians are banging away. But the Turks have brought up heavy reinforcements, apparently fearing a big attack from this flank. Fragments of shell and shrapnel-pellets are whizzing amongst us even though we have splendid cover. The shells from both sides are screaming overhead making hell's own row among these rocks. On the crests behind us, hundreds of

Bedouins are looking on.

As they stand up or squat down, we can swear their white-and-black shrouded figures are signals to Turkish observation posts across the valley … The battle is raging—we wonder how things are going; apart from the rolling thunder we only hear rumours, whispers. We know that around the Tel el Khuweilfe fortifications the fighting is desperate, the 1st Light Horse Brigade, the New Zealanders and the Camel Corps have gone in on foot to help the infantry. Their horses had been forty hours without water and had to be sent back to Beersheba where they went mad immediately they smelt the water-troughs. Rumour is varied as to the struggles around the great redoubt systems stretching right to Gaza, but Khuweilfe is in everybody's mind for it holds water, water, water!

Apart from us knowing that the navy and land batteries are putting Gaza under such a terrific bombardment as the Turks never dreamed of even in their worst experiences of Gallipoli, we have heard only one wisp of Gaza news. The Scotties stormed Umbrella Hill. Their first wave was blown up by land mines, but successive waves rolled on and took the hill with bomb and bayonet. The 54th Division with their six tanks are experiencing bitter fighting.

A few hours later—We were all sitting down hungrily eating bully-beef and biscuits when a screeching hail of iron and splintered rock smashed throughout the regiment. Men were down—horses down, others rearing—one poor brute balanced on its head for seconds before it collapsed and rolled down the hill. Crash-crash! crash-crash! crash-crash! Jagged iron and lead and shattered rock and more men down—more horses, some stone-dead already, others writhing for foothold on the steep, rocky hillside, while other animals not wounded to death just stood still, dumbly wondering why warm sunlight had been transformed into hell. We rushed to help the wounded—to catch the horses whose masters were down—seeking then quickly what poor shelter the hill could give. For suddenly what had been good shelter was now none at all. The Austrians had brought up something doubly new in howitzers, the shells from a dizzy height came straight down—their trajectory would sheer a precipice and they burst behind the hill yet in against the very rocks behind which we were sheltering. Each shell burst with a double crash as if two shells joined—one half exploding from the other and kicking backwards. Instead of the big iron shrapnel-case shrieking empty to the ground it too was filled with explosive and burst into a thousand pieces. One horse had his neck cut clean in halves by a single fragment. Just a few shells—direct hits—but what a mess they made! For the remainder of the afternoon we were helping the wounded down from the rocks, seeking cover somehow, somewhere, anywhere for the unwounded horses, shooting the poor brutes that were too badly hurt to save. I saw men nearly cry to-day—we love our horses. Again and again the Austrians swung their guns on us, plastering our battery or any wisp of dust that might betray horsemen seeking cover. At such times the air above was a splitting terror of doubly bursting shells. I heard odd rifle-shots and knew that some maddened trooper was taking his revenge for his horse, for on the rocky crags behind the Bedouins like black vultures had stood all day, watching us. Looking up this our hill this evening, sombre under the setting sun, with its blood-stained rocks and dead horses, it makes us feel what a miserable thing war is. And we hate the Bedouin. And know a fierce resentment against the Turk who waits for us to pant up those great hills opposite and dig him out of his snug redoubts with the bayonet. Colonel Cameron is downcast over this morning's sudden tragedy.

Next day—Last night we lined the big hill-top expecting a Turkish attack. No such luck. This morning at three o'clock we left a few men behind to deceive the Turk into believing that we were still there while the whole brigade rode back towards Beersheba, for something is doing for us. As we climbed ever down from among those black Judean hill-tops we could see far away even towards Gaza. It was like what the world might have been making, myriads of dancing flames in the darkness. Only a few miles away the guns roared in echoing reverberations upon the fortifications of Tel el Khuweilfe. A few miles farther, was a terrific pounding upon Tel Sheria and we could plainly hear the close thunder rolling over Hareira. From the distant Atawineh redoubts came the sullen muttering of a storm, with sheer surmise as to what was happening around Ali Muntar and the strongholds of Gaza.

The sky for miles around Hareira was pierced by lightning flashes that illuminated hill-top after hill-top. Balls of flame seemed actually revelling in the blackness, the sky was streaked with coloured lights. Then we climbed down into the black valley and seemed to breathe in thunder. As we neared Beersheba the roar of concentrated rifle and machine-gun fire grew. Its menacing harshness entered a man's toes and shivered right up to the roots of his hair.

3 p.m.—We are close by Beersheba. What a hive of activity the town is, especially the great wells where gangs of Engineers worked in almost frenzied energy around pumping-plants, erecting great lines of troughs. All the wells are a mass of thirst-maddened horses. The 8th Light Horse have been fighting with no rations, their horses forty hours without water. The New Zealanders and the 1st Light Horse Brigade are fighting on foot, their perishing horses are frantic around the wells.

The 53rd Division and the Camel Brigade are fighting bitterly around Tel el Khuweilfe for that place is an important road junction seven miles west of Dharayet. But above all it holds the water without which great bodies of troops cannot push forward to the next phase of the fight. Some of the 7th have just hurried off to lend a hand. We wonder are we all going! … The brigade has just got orders to move out. Vanish our dreams of a sleep to-night.

Sundown—We hear startling, splendid news. Chetwode's division by a grand assault have smashed the Turkish 7th Army at Kauwakah. I think Chetwode's men are the 10th, 60th, and 74th Infantry Divisions and he has the loan of the Yeomanry Mounted Division from Chauvel. We are delighted that the Heights have fallen.

…News comes fast. The 4th Light Horse Brigade are into it side by side with the Londoners at Khuweilfe. The fighting there has been bitter for days past. The Turks are steadily reinforcing their losses—the casualties are severe on both sides. There has been bomb and bayonet fighting, galloping of positions and machine-gun fighting at forty yards range.

Morry's horse was wounded yesterday. Morry has had to go back to the Mobile Column and on the eve of such great events, too! The section is dashed sorry.

Men of the 5th Light Horse on alert near Ghoraniye.

46

November 7th (I think) — We travelled all through the night, a night filled with the roar of guns — flamed by lightning — harsh with crashing waves of sound ceaseless and frightening: the air saturated with wonder and surmise. The creak of wheeled columns and the hooves of our own brigade murmuring all through the night; choking dust so dense that columns only yards apart could hear but not see; rumbling down into black gullies with grating shriek of interlocked wheels; sparks from shod hooves as chains tautened to the straining teams labouring the guns up, up, up again into the fog of dust; knowing we were passing fighting-ground by the smell of the dead — so Australia's mounted army rode into the Plain of the Philistines, half paralysed by the longing for sleep. Early this morning we halted in this Wadi Ghuzaley; its steep banks are sheltering us now against the screeching Turkish shells. The remnant of a Welsh battalion is here: yesterday they were badly cut up in a Turkish counterattack. We are all waiting now for our own turn to attack. The waiting is nerve-racking. There are such a lot of dead men lying about too.

...Great news has come through — the 10th Division has rushed the Hareira redoubt. The great mound of Tel El Sheria has fallen to the 60th Division, the 5th Mounted Brigade, the 3rd and the 4th Light Horse Brigades. Hurrah! Mounted troops are pouring into the plain. Von Kressenstein is rushing columns of motor-lorries packed with men to Tel el Khuweilfe, for there lies the water for an army Our nerves are raw, and we keenly feel the loss of tobacco. Men are smoking horse manure, tibbin, dried tea leaves, grass — anything. So long diary — "Mou-nt!"

The day after — very early morning — The brigade hurried off down the wadi until its banks opened out into rolling downs country with the blue range of Shephalah on our right and the formidable positions of Sharia and Hareira on our left. The Turk is fighting magnificently. As position after position is taken its survivors fall back on prepared fortifications behind and reinforced by fresh troops fight with fatalistic determination. But now we have broken through their centre. From miles away in three directions they pelted shells at us but we had splendid cover. Then as General Chaytor came along we cantered out to some chalk hills then in full race galloped down a column of retreating transport-wagons. The place is Kh Urn El Bakh. The Turks retreated too fast to blow up the great stacks of ammunition and baggage. From a hill-top we watched a grand sight on the plain. There was Sharia spread before us and near by the red station houses of Amiedat, the tents of a far-flung Turkish camp with redoubt after redoubt all writhing infernos under bursting shells. Advancing steadily just behind a spouting barrage were long lines of our own infantry. Ahead of them were the Turks running back with away on their right a rolling cloud of dust in which thundered the 1st Light Horse Brigade galloping to cut them off; all troops moving under clouds of shrapnel. We watched until dust and smoke and shattered earth enveloped men and horses, infantry and artillery, friend and foe.

Then presently we saw the brigade re-form from out the mêlée and breaking into a gallop race for Amiedat. We held our breath for we could see trains puffing at the station and we knew the Turks would fight frantically at sight of this enveloping charge ending their getaway. But the great charge at Beersheba seems to have terrified the Turks of "the mad Australians," as the prisoners describe us. The 1st Brigade swept on under a tornado of shells and machine-gun fire until they disappeared among the buildings and huge supply-dumps—we knew they must have had a great capture. Excitement was all over that vast Plain of Philistia.

Men are remarking how the Turk fights to the very last charge, until the pounding hooves are upon him, then he drops his rifle and runs screaming; while the Austrian artillerymen and German machine-gun teams often fight their guns until they are bayoneted.

We hear now that a Yeomanry general pleaded with Chauvel before Beersheba for the honour of the charge that took the town. The Yeomanry are armed with long swords, a terrible weapon on horseback at close quarters, whereas the Light Horse have no cavalry weapons at all.

"Brigade—Mou-nt!" And away we went full gallop across the plain; but the Turks were being handled so severely elsewhere that only the guns from the Judean hills could spare time to shell us. In artillery formation we spread out into a walk, sparing our precious horses, watching the shells fall in the open ground spaces among the troops and miss and miss and miss. Our regiment then occupied Tel Abu Dilakb at the gallop. We were miles closer to Gaza now, sinking a wedge right in behind the doomed city whose hill-tops were spouting columns of earth from the shells. Up from Amiedat trudged columns of Turkish prisoners with hundreds of transport wagons drawn by queer-looking oxen. We grinned at some German officers, for they looked very hot and dusty and annoyed.

Two guns in particular four miles ahead, had been harassing our regiment. So when the Old Brig. came along at a hand gallop, pipe alight, we formed into artillery formation and galloped straight for the guns. How they plastered us with shells! By Jove it was a grand gallop with the horses reefing for their heads, three blood boiling miles with the screeching air-blast of the shells in our faces, then a thrilling loneliness as A Squadron galloped on alone—into a burst of machine-gun fire, vicious rattle of rifles. We saw lines of infantry lying behind the guns—every man stared—numbers laughed as we pressed knees into our horses—we galloped over a low hill down into a gully, our horses on their haunches as the major shrieked "Halt!" We peered over the sheltering bank at a flat running for six hundred yards in full view of the German machine-guns—then a smaller gully and just beyond its farthest bank the guns waited—relays of Turkish infantry lying waiting too!

Our first troop faced the gully bank, then at a signal dug in their spurs. The horses plunged up—we caught our breaths at the hail of shrapnel as their tails swept up over the bank! The gunners had the point-blank range to a breath. Then another troop—three sections at a time spurred over at erratic distances apart—the shells and bullets churning a lane of whistling earth out of the gully lip, but the next troop went over just a little farther up or down the bank and the Austrians lost seconds in swerving the guns. Our turn came. The first three sections jammed knee to knee against the bank—our hearts thumping—minds racing with a variant dodge to fluster the gunners. We waited (My God! how breathlessly) until two shells exploded crash-crash!—we spurred up into the cloud of smoking dust, horses pawing the bank, and were racing for our lives. Crash-crash! a blinding flame tore the ground in a whirlwind of dust and fumes, tearing screech of high-pressure bullets—we strained hold of our maddened horses then, Christ! a precipitous gully in front—the awful sensation of a void—a breakneck swerve—a crossing came thank God—scream of shells passing feet away—hot breaths of bullets thicker and thicker—horses berserk—the certainty that horse and man must crash before reaching the shelter now so near, then another wild swerve and we were down in among panting horses and the men who had gone before. Pools of blood were forming there; an old chestnut horse I had often ridden held its head pathetically, bleeding from the nose. Lieutenant Webster lay dying; his men liked him very much. Others were down, but our casualties as almost always were extraordinarily light. I don't think any regiment in the world would escape annihilation if it tackled two Australian guns supported by machine-guns and riflemen—and we were only a squadron!

We lined the ridge and opened out on the Turks. They would not stand—blazing with their guns as they steadily retired before us while reinforcements appeared behind them rushing up more machine-guns. It was almost sundown and presently we had pressed on a long way ahead of our horses but the Turks refused to stand, just fighting us yard for yard, retiring as we advanced, fighting yet luring us on most cunningly. We fixed bayonets and crept forward ready to charge the shadowy figures immediately we could get close enough, but in the heat of anticipation the order "Retire!" was signalled from the invisible brigade behind.

We galloped back and presently formed outposts when to our intense surprise most of us lay down by our horses to sleep. It appears the Anzac Mounted Division had been ordered to await until Tel el Khuweilfe had fallen and the Australian Mounted Division could catch up. We had had no water since leaving Beersheba.

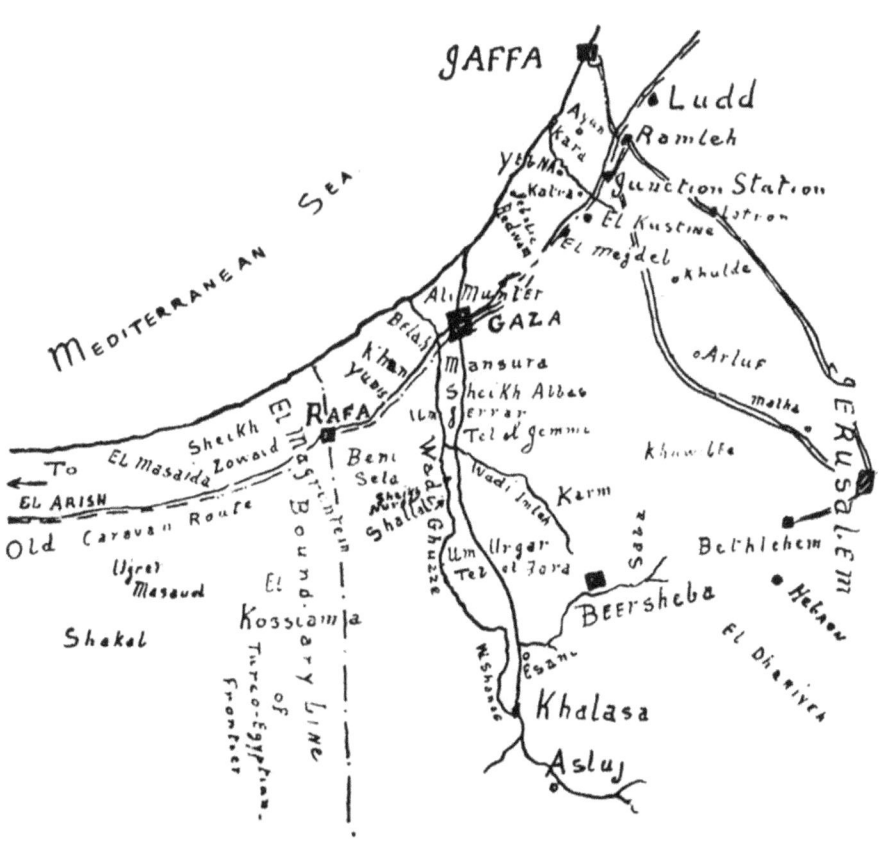

47

November. Early morning—Yesterday was a wonderful day, from dawn until dark. Long before the east grew leaden the whole brigade, all unseen, was sitting down by its horses' hooves, just holding the bridle while the poor old neddies stood motionless, some with one eye open but each watching its master, its muzzle within six inches of his face. They had now gone thirty-six fighting hours without water. For ourselves, we still had our Iron Rations but were craving for tobacco. Dawn came in waves of colour that bathed the Judean hills Caesar's purple and turned pink the hills of Beersheba. Then the Plain of Philistia brightened and presently the Gaza hills were outlined against the western horizon. Most of the 6th officers were observing from a ridge in front. We were to take those two guns this morning; we would gallop into no big trap in daylight.

The Anzac Mounted Division was now ten miles right behind the enemy's front-line. It was rather thrilling, miles in behind a great army fighting furiously for its existence. Then news came that the forts of Khuweilfe had fallen! That news thrilled the whole brigade—not only because of victory, but there was water, water, water!

We grouped on the little rises and gazed back towards the Beersheba hills. As far as the eye could reach were rapidly ascending columns of blood-red dust. Thirty thousand mounted men were pouring into the Philistine Plain. We could see only those whirlwind columns but the knowledge that they were our own mounted men thrilled a man's very heart.

With what desperation the Turks must have watched those ominous clouds! Someone called out: "I wonder if Chauvel is watching from Beersheba."

That one day must have been worth a lifetime to Chauvel. It saw him in victorious command of the greatest body of cavalry in the history of war. But the awful anxiety concerning the fighting still to come—for water! With camels and transport there were a hundred thousand animals to water, and at the close of summer the wells of Palestine run low!

Suddenly Major Bolingbroke shouted to Captain Fitzpatrick: "The guns are escaping!"

Instantly the squadron was in the saddle, with a clatter of hooves we were plunging up over the skyline then away. It was a wild gallop, up hills, down hills—leaping gullies—taking sun-cracked crevices in our stride—seeing in plain view two miles away the guns toiling along a road. Some of the boys whipped off their hats and laughingly smacked their neddies' rumps, for we hated using spurs on the poor thirsty beggars.

How different to yesterday! The Turks then were securely sheltered with battalions behind them; we were one hundred men in the open. Now both sides were in the open. We could see the infantry escort far ahead of the guns hurrying their machine-guns away.

Nearer we galloped, past boxes of shell ammunition thrown hastily from the wagons—the gunners were throwing off more ammunition, thrashing their poor little teams. Along that road, a column had retreated in haste the night before. There lay hundreds of infantry packs, gas-masks, gear, rifles, baggage, bayonets all littered along the roadside.

One of the guns halted in desperation and swerved the ugly muzzle straight at us. But the squadron dug in the spurs, stood in the stirrups, waved bayonets, and roared! It was enough! the artillerymen lost their heads, a mounted officer struck in mad exasperation, then the officers leaned over their horses' necks and galloped off along the road, the artillerymen ran for their lives. The squadron roared laughing. As we galloped past the first gun I saw a blood-stained Turk lying beside it.

No shells screeched from the guns' black muzzles now. Squat, solid brutes they looked, glumly silent.

We swung off our saddles, slung the reins to the horse holders, ran out in front and knelt down: *crack-crack-crack-crack-crack-crack—tut-tut-tut-tut-tutututut-tut-tut*—and the panting artillerymen must have coughed up their very hearts, for by Jove they ran until they fell! How great it was—spurts of dust kicking around their desperately moving legs. They had sweated to give us hell yesterday. But what an incomparable difference in the aim of the Light Horse trooper! The only artillerymen who escaped were those who threw up their arms, and a flying group who leapt exhausted into a ravine; they lay there panting when we galloped up.

The regiment was coming away behind with the 6th and 7th Regiments spread on either hand, a long line of the dreaded "Felt Hats." Away behind again as a dust-storm entering Palestine thundered the Anzac ten thousand. The sky was azure blue, glinting on buzzing machines—shrapnel-puffs dissolved in lazy threads of gossamer. The irrepressible lark trilled high up, shells screamed far over the plain, red earth spouted skywards. The squadron galloped down the tell-tale road littered with the wreckage of an army. That was a great gallop, in the middle of it we plunged down into a deep wadi through a pool of clear water! How longingly our horses thrust down eager mouths! It hurt us more to force them to splash through. Then up over a low rise and a canter across bare cultivation land, to race down into a valley and suddenly gallop into dustless country.

On a hill was a commanding stone house, squat and square, built just below it a garden of fruit-trees and among them a huge stone well with a wooden water-wheel and cement tank. The enemy were in great force just ahead—we had far outstripped the brigade—we pulled up and the horses rushed the well. The conveniences made watering quick and easy.

Bang-whee-eee-eezz-crash! and shrapnel pellets lashed the orchard-trees. But the Turkish guns could not stop us watering. Then we trotted off to the shelter of a little white hill close by. A shell burst above the house roof—out scurried the Arabs like frightened rats making for the shelter of the hills. Soon, a squadron of the 7th cantered up and away we went reinforced and quick and lively were greeted by machine-gun and rifle fire. Cantering over a ridge we gazed at a long line of Turkish infantry advancing steadily towards us.

"Dismount for Ac-tion!"

We lined the ridge—the Turks were almost on us, their numbers rapidly increasing. We simply stood and fired as fast as we could work our rifle bolts—the Hotchkiss guns shrilling one long, hot scream. Suddenly a large body of Turks appeared encircling our right—hails of bullets came from two directions—we sprang on our horses only when we could plainly see the savage eyes of the leading Turks. What a gallop! Over gently sloping ground—not a leaf of cover—ten times our number furiously blazing at us. For more than a mile those vicious pests whizzed past ricocheting off the ground, flipping off a man's hat, thudding through a bandolier. How the old neddies legged it out! I began to think we would never get out of range of those high-powered Turkish rifles. At last we swerved in behind a hill, hurriedly picked our possies and waited for the Turks to come on again. They came, steadily, irresistibly. We blazed at them—we fired and fired at the splendid targets marching steadily on. They got right to the foot of the hill, their thinning rank came steadily on but other ranks were hurrying past our left and right, we even heard the thump of their feet as suddenly they doubled in on us; so close they got we saw their panting mouths as they howled when we leapt for the saddles and away. It was a hell-for-leather go, bending over the neddies' necks, laughing with the wind in our teeth. We dismounted behind another little hill a half-mile farther back and running up near the skyline lay down and waited again, filling our emptying bandoliers with the spare bandoliers collared around the horses' necks. The Turks came steadily on—fine-looking chaps in their blue-grey uniforms. They must have been sweating hot though. We blazed into them until our rifle-bolts were blistering hot before we rushed the horses and were away again.

And so we fought until midday, when the brigade came cantering up. That little fight was splendidly managed. The two squadrons had got into a very tight corner, we fought and retired, to fight again with very little loss to ourselves and as was proved afterwards we had seriously delayed a body of reinforcements whose strength was twelve to our one. The brigade took up a position along the rugged banks of the Wadi Hesi. The Turks halted, our glasses showed they were awaiting strong reinforcements hurrying up behind. In the midst of the excitement came news that Gaza had fallen and that the Turkish Main Army was now cut in two and in full retreat, putting up desperate rear- and flank-guard battles to allow their retreating forces a chance to pour down the Ramleh, Jaffa, and Jerusalem roads. It was stirring news, but just fronting us— They looked simply grand, those chaps; lines of big grey-clad men coming to push us off the face of the earth. We were jolly anxious for we were one little thin line against thousands, and we could tell by their swinging march that these Turkish troops were fresh. Our little guns opened out at point-blank range, their support was inexpressibly heartening. Then broke harshly the machine-gun and rifle fire along the wadi banks—but the blue-grey lines came steadily on. In the fury of it all a squadron came galloping up—late. It took its place in the line in a shower of gravel and dust—the men sprang off as the horse-holders snatched the reins and wheeled galloping away—the men ran forward and got their Hotchkiss guns into action right in the face of the Turks. It was a great little piece of work at blood-boiling pressure.

The first Turkish line melted away right in front of our centre. The second line came on, dwindled, the third line came on. Hesitated, turned and retired steadily. They re-formed, and came again. Again they were wiped out. The remnants retired and lay down and blazed away at us, awaiting apparently inexhaustible reinforcements. We fired back—it was wonderful shooting. Hearing a tremendous buzz in the sky I looked up and there flew a great sight—twenty-eight roaring birds going to bomb the Turks along the Jerusalem road! Soon we heard explosions that thundered among the Shephalah. What hell those close-packed battalions must have got, hedged in by the cliffy hills! Presently, back they flew again in clockwork formation, very business-like, roaring through the sky. An hour later they returned, and again we listened to detonation after detonation as they helped spread demoralization among a retreating army. In late afternoon, we suddenly stared. As far as I could see, for a length of five miles, coming across the open country, was a line of grey-clad men. They came steadily on. Two hundred yards behind them, came another long line. When in full view, behind them marched another. Then we got right into it and I could see no more except with a throb of thankfulness when two regiments of the 5th Mounted Brigade, their gun wheels spinning, came thundering up behind us. They ran into the line just in time to be heavily attacked. The Yeomanry fought so heartily we claimed them as brothers from that afternoon. It was a great fight right out in the open, the Turks determined they would advance, we determined they would not. Our men steadied down to that cool, deadly firing so noticeable when we are hard pressed. How differently we shoot to the Turks in similar circumstances! They suffered terribly—came on again and again, brave men indeed! Numbers of them fell almost at our rifle-muzzles. At sunset they were done. Their remnants lay down, still firing. Miles ahead along the road blazed fires like streets of houses where the 'plane bombs had ignited the Turkish supply-dumps. Presently, fires in soaring flames of oil leapt up; then a series of rocking explosions that sprayed the sky with fireworks. The Turks were blowing up ammunition dumps all through the night. Late, we retired half a mile to where we could get a little water. Then went on outpost duty.

A halt on the march. The 1st Light Horse regiment near Esdud, Palestine.

48

Next morning (which was yesterday) at grey dawn we mounted. The Gaza garrisons were retreating down the Jaffa road; we must fight our way through miles of Turkish flank-guards and try and cut the garrisons off. Under shrapnel we rode out, our bellies empty, but we almost cried for our poor horses. Nothing to eat for forty-eight hours, and again no water. We rode straight for the tail of the Turkish rear-guard and quickly saw numerous movements of troops, their reinforcements hurrying up in motor-lorries. As the sun rose, so grew momentarily plainer along the road ascending columns of smoke. Far across the countryside were other rolling clouds where the retreating army was burning stores. Frequently an earth-shaking explosion advertised another ammunition dump gone sky-high while much smaller concussions denoted the blowing up of precious wells. The rear-guard tore their shrapnel into us—they worked their guns viciously. Bert and I had a narrow shave, a shell crashed just above us and down thumped the horse in front, Dick Rutledge with him, a bullet in his leg. A quarter-mile farther on Bert exclaimed: "Hell 'n' tommy! my horse is hit!" There was a pellet clean through his saddle but the impact had flattened the bullet and it only just penetrated the horse's side. As I dismounted I cursed everything, for blood was streaming down my poor little mare's leg. But she had only been hit in the rump, thank goodness. Still under shell-fire, we trotted across a flat and past some transport wagons, the bullocks dead beside them. Then came a line of wagons, numbers smashed from shell-fire, the poor bullocks dead or dying. Then over a hill we rode on a smoking aerodrome. Hundreds of cases of bombs and hundreds of wicked-looking aerial torpedoes were lying about. Numerous torpedoes stood the height of a man, and with their mysterious gear they looked formidable enough to sink a battleship, let alone blow unprotected men and horses sky-high. A taube smoked in ruins while the pumping-plant of a nearby well had only just been blown up. We put the spurs in and galloped into a large camp, the tents still standing, men left behind hurriedly packing. Numbers surrendered. Farther on was a dead bullock half skinned, showing how hasty were the breakfast-less ones.

We cantered on and quickly rode through other camps, abandoned wagons everywhere, hundreds upon hundreds of mules, camels and oxen lying dead of bomb, shell, bullet, or exhaustion. We were expecting anything to happen. Two guns in particular had been roaring in our very ears for some time past. Everywhere was thickly scattered the abandoned equipment of an army. Then we rode straight on two squat howitzers, the beastly muzzles still smoking—the gunners flying on horseback. Two sections of our screen went straight at them. Three men mounted on the fleetest horses looked like getting away. From a low hill, we saw Corporal Cox far outdistance his section, then jumping from the saddle kneel down and—*crack!*—down crashed a horse and rider. Cox leapt on again and galloped to again jump off—crack! again a man and horse sprawled in the dust and Cox, we could just see him now, was after the last man, an Austrian officer he turned out to be. Cox fired and the horse crashed down but the rider rolled behind a rock and with an artillery rifle shot back at Cox. He missed—Cox didn't!

It was rattling good shooting on the corporal's part. I'll bet no Turk could have done the same to three of our chaps.

Then we cantered down into a larger camp where hundreds of Turks were frenziedly packing. They stared up from their task in amazement and fear; some fell to their knees as we galloped down amongst the tents. Only a few showed fight. I laughed at a natty Turkish officer furiously striding along a line of half-loaded camels, swishing his cane. He was caught nicely. The camels gazed calmly on. Broken-down wagons littered the road in ever increasing numbers, just as shell or bomb had found the poor bullocks. Other wagons all fully loaded, had stopped just where the bullocks fell dead of exhaustion. Crumpled motor-lorries lay in amongst the wagons.

Then C Troop was ordered back to guard the guns lest reinforcements should turn our flank and train the weapons on our rear. On our way back, we met hordes of "vultures," hundreds upon hundreds of Bedouin men, women and children driving donkeys and camels, all horribly eager to loot the Turkish camps. Numbers of these Arab cut-throats carried sacks of little flat loaves of brown Turkish bread, looted from the still warm ovens. We hungrily commandeered some of that bread. The Arabs snarled at us. We could have shot them with far greater pleasure than we shoot Turks.

For several hours we stood by the guns, then were delighted at sight of our Transport Column hurrying up as fast as the horses could go. Quickly we stopped a wagon seeking our horse-fodder while the column pressed on to overtake the brigade. Our horses wired into their fodder but we could find them nothing to drink, About four o'clock we left a dozen men in charge of the guns, then pressed on after the brigade. They must have moved very rapidly for at dusk we had not caught up. Our horses soon were stepping in and out amongst the confusion of Turkish baggage, gear and equipment, and occasionally a blood-stained Turk. At dark we rode on a large Arab town, and a most cruel sight. Hundreds upon hundreds of Turkish wagons in jumbled masses of wreckage and confusion. Some were piled upon the other where bomb or shell had blown them. Streets, roads, houses, cactus hedges were littered with fragments of wheels, wagons, limbs and entrails of bullocks. But hundreds of wagons were quite unharmed, their horses and bullocks and mules still yoked to the wagons, in a pitiable state. Those that had been killed outright by shell, bullet, or machine-gun were the happy ones. Many had dropped dead toiling under the knout until their hearts burst. Others were still alive, moaning with the piled-up wagons on top of them; others were being slowly strangled by the yoke and weight of the loaded wagons. A few lucky ones, where shells had smashed the wagon-shafts, had broken away, but a great number of the live ones were slowly dying of wounds, exhaustion, hunger, thirst, and strangulation. May I never see such a sight in any Australian town!

Coming and going from the town to the wagons were swarms of Arabs, men, women, and children, staggering under loads of loot, panting, struggling, the sweat even at eventide pouring down their swarthy faces in their greed to get all away before more distant villagers should rush the spoil. I felt quite sick watching the greedy women, hysterical, shrieking, almost crying in their excitement as they clawed the contents of the wagons not knowing in their envy and fear which of the stuff to take. They didn't think to end the misery of the poor moaning beasts upon whose mangled bodies they climbed up into the wagons. Among indescribable wreckage and abandonment of goods, over hundreds of yards of ground were scattered the records of a Turkish division, sheafs of private letters trodden like snow under the dirty brown feet. Countless rifles, stacks of cases of ammunition, wagons loaded to the brim with shiny brass bombs, farriers' gear, saddlery, armourers' gear, wagons loaded with doctors' gear, with officers' gear, provision, confusion indescribable; the baggage of an army in full retreat.

Among the huge litter lay dead Turks, their dusty faces trodden on by the feet of men, women, and children. The wheel of a wagon had been blown away so that one end sagged nearer the ground. Right under the corner lay huddled three Turks used as a footstool by women and children whose blood-stained feet were covered with dust. We wondered if the Bedouins had troubled to put the wounded quickly out of misery in their hurry for loot!

Feeling rather sick we rode on following the same trail, broken wagons, dead beasts, Turkish clothing, smashed men, greatcoats, baggage.

Long after dark we camped and early this morning made for a big Assyrian town and located its well, two hundred feet deep. Our horses were perishing. We have had a meal ourselves and a quart of blessed tea. It is twelve o'clock now. We got plenty of grain from a Turkish wagon and our neddies have eaten until they can eat no more.

The mounted men are pushing the Turk fast towards Jerusalem. Jerusalem, city of peace and hope! Its byways now are roads of hell. Two miles ahead of us is a roar of rifle and machine-gun fire; it spreads far away to the left towards the sea, we can hear its faint echoes far to the right amongst the grim Judean foot-hills. How the robber bands must be listening in their caves! Many robber bands inhabit the fastnesses of the Shephalah. I wonder if they watch, in an agony of longing, for a chance to swoop down on the loot! They who had once levied tribute on Palestine now line their hills like caged eagles.

All ahead of us are slow drifting clouds, for the shrapnel-puffs are so thick that they have merged into big white clouds. The roar of guns is rocking the air. The Turkish rear- and flank-guards are fighting desperately—grandly.

The country is changing rapidly. On every little hill nestles a thickly populated village surrounded by cactus hedges and trees. Some in the distance are towns. The downs country is all open between villages, much of it under intense cultivation.

The magnitude of these operations has long since got out of my focus but the officers say that as the Desert Mounted Corps sweeps on right up Palestine, the Anzac Mounted Division is spread from the coast inland with its right joining the Australian Mounted Division which spreads out with its right touching the foothills of the Shephalah. Thus Australians and New Zealanders stretch right across Palestine from the coast to the Judean hills, riding in a "face" right up the country, packing the Turks along every road back towards Jaffa and Jerusalem. The Yeomanry Mounted Division follow behind the Anzacs and the Imperial Camel Corps ride behind the Australians.

We are moving off again and thank Heaven I have got this diary up to date. It worries me.

49

November. About 2 p.m.—The brigade charged in a whirlwind gallop under a furious fire and smashed through the Turkish rear-guard, taking fourteen hundred prisoners and thirteen guns. We hear that the 1st Brigade farther along the line also took over a thousand prisoners and some guns. I don't know our casualties but we see that for every Australian killed there are twenty dead Turks. One troop of the 7th stopped a direct hit—men and horses blown up together. The 6th Regiment experienced stiff fighting but as I wasn't with the crowd yesterday afternoon and last night, I can't write about it. The Anzacs now have to hurry on alone, the Australians and Yeomanry are held up through want of water.

We are camped in a very large village trying desperately to get sufficient water for our horses. The Turkish prisoners moaned ceaselessly last night for water. The village inhabitants are swarming the wells for the Turks had taken possession. The wells are almost dry, the water is dribbling in very slowly, pathetically so, for the two thousand horses of a brigade … I would have liked to have been in the gallop through El Huleikat, the headquarters of Von Kressenstein, G.O.C. 8th Army. It was hours of the wildest excitement and a stampede of galloping charges with the climax, disaster for the Turks … We only get wisps of news about the doings of the other brigades let alone divisional and army corps events. We know that the Australians had a desperate fight before they took Huj, the Turkish Army headquarters.

The Australians are enthusiastic over a charge by two Yeomanry regiments (the Warwicks and Worcesters) who galloped up at a critical moment and with drawn swords charged straight into the mouths of eleven guns, sabreing the gunners then galloping down the supporting machine-guns. It must have been as wonderful a charge as the Australians say it was. The Yeomanry will have their tails up now. By Jove, they've earned it. Eleven guns! supported by infantry and machine-guns!

The 3rd Light Horse Brigade were wishing for swords—they broke through the Turkish flank-guard, killed hundreds and took a long column of prisoners. If they had had swords the men swear that they could have cut the retreating columns up in such a way that the division would have taken thousands of prisoners. If we could only keep on like this the war would soon end.

This village like the others in sight is perched on a low hill-top, surrounded by prickly-pear hedges enclosing fruit-trees. In the centre are the square-built mud and stone houses with their thatched roofs. A German officers' camp here abandoned a score of wagons, violins strewn about among littered music, gas-masks, clothes, cartridges, greatcoats, and ladies' photographs.

But there are a dozen great hogsheads rolled against the pear. The mean swine had just time to pull out the bungs as our chaps galloped up. Their whole camp stinks of beer. In my hurry to write up this morning's notes I forgot the most important thing of all. Yesterday morning in one Turkish camp we galloped into were cases of tobacco and cigarettes which we swarmed as ants swarm honeycomb. The Turks just gazed, their hands held above their heads, staring as if convinced we are really as mad as their officers tell them we are. In a twinkling every man was sampling Turkish cigarettes or puffing from his long hungry pipe—It was great.

The country is becoming more attractive. Every hill has its village, much larger and cleaner and more prosperous than any we have seen before. Perched on the low hills surrounded by their cactus hedges, with the roofs and domes of the queer eastern buildings peeping above the trees, and a white stone mosque crowning all. Numbers of them have quite a "Sinbad the Sailor" appearance. The green fields in between are intensely cultivated. Two miles ahead of us is a quaintly pretty little town; all white walls, red roofs and the greenest of trees. The inhabitants are prosperous looking, numbers of them are actually clean. The young girls at times are surprisingly pretty, tinted browns and whites with flashing black eyes. Shrapnel is bursting over towns and villages—we hear the roar of machine-gun and rifle fire.

I hope war never comes to Australia! If it does, may we have arms efficient enough to keep them from shelling our own country towns. The English don't shell the towns here, it is the Turks and Germans as they evacuate.

We are taking but passing interest in the shells now. Water is our ever present trouble ... We hear that the great Arab town of Medjel has been occupied ... The Turkish counter-attacks are being beaten off but still they come again—and again—and again ... We are being shelled from El Kustine ... Our fellows have beaten back the Turkish waves striving to retake Burier and Sirnsin. Away on the left, the New Zealanders are meeting furious opposition ... Rumour comes that the 1st Light Horse Brigade has captured Es Dud.

Our Bible enthusiast is bubbling over, for Es Dud is the ancient Ashdod of the Old Testament, the Ashdod of the Philistines Cox's brigade just saved the bridge at Jisr Esdud and are defending it by a bridgehead.

Our troops have galloped over Ascalon, and the Bible enthusiast has balanced himself on a German beer-barrel (empty) and got this off his chest: "O man, savage, ferocious, what desolation hast thou wrought on the earth! They have stretched out upon Ascalon the line of confusion and the stones of emptiness. Thorns have come up in her places and brambles in the fortresses thereof and it is a habitation of dragons and a covert for owls."

We cheered him: he bowed and gave us a lecture. It appears that Gaza and Ascalon, Ashdod, Gath and Ekron were five capital cities. Some old prophet cursed them and the prophecy came true. Ashdod the Proud defied the siege of Psammetichus for twenty-nine years. An old Crusader castle, ruined like the city, still guards the forlorn harbour. At Ascalon city, Herod was born. Gaza is the only city of them left. A shell has just burst above the Bible enthusiast and he ended his lecture abruptly—five shrapnel-bullets whizzed right through the barrel.

…The Australians have taken some enormous convoys. Firing is rolling right up from the coast to the Judean hills. The fellows are cheery as anything, even if we are covered in septic sores and awfully sleepy …

We are in desperate straits for want of water … We hear that the New Zealanders' horses have gone seventy hours without water. Two Australian brigades have gone sixty hours without. The Yeomanry have done a perish, too … The Cameleers are being severely counter-attacked. Their "ships of the desert" look grotesque lumbering into action while modern armoured cars roar past then spiting flame from machine-guns … Both Morry and Stan have to go back to the Mobile Column—their horses have been again wounded. What a shame—after all these months to miss phases in the most interesting battles of history … Everyone is delighted with the success of the operations. General Allenby has made good.

Next morning—To our interested surprise yesterday evening, the 52nd Infantry Division marched up and took over from us. They must have flown! We rode to the left for about seven miles and struck the beach where there is plenty of water. Our horses are deadbeat. Our troops have been wonderfully quick in cutting the railway lines and blocking the roads. We have cut the Turkish army in half but the other half have escaped and is fighting desperately. Our organization has functioned splendidly. Fodder and rations, despite the wildfire movements and huge area of country fought over, have been kept up to us wonderfully, all except the most important item—tobacco.

…How strangely peaceful this morning seems! Our horses, some lying down, others standing up, are quietly resting after exhaustion. The sky is heavenly blue. The sea peaceful. The air sweet—not one speck of dust. Miles away to our front and right comes the hum of war. Tired though we were, we went to the trouble of scraping holes in the sand and burying a few dead Turks who were polluting the beach. And now we are lying on the sands, dozing. We may have to mount and hurry out again at any moment, but for just now we are at peace. And I waste the precious seconds writing up this old diary!

November—Hamane. Around us are sand dunes, farther inland are the prickly-pear hedges hemming in the orchards, the olive, almond, and orange groves. The inhabitants guard their trees as if every orange were made of gold. We are again disgusted with the lies of the papers, stressing the wretched inhabitants of Palestine as starving with hunger. Such possibly may be true of the country below Jerusalem, certainly not of the people we have seen so far, except those unlucky villages and colonies whose inhabitants have been of a different religion to the Turks. The men, women, and children we have so far seen in this agricultural Palestine are all fat and healthy, have plenty to eat, and good clothes. Their flocks are fat and large, so far as flocks go in this thickly populated area.

Again, the papers write glowingly of how the "Liberators" are welcomed with open arms and tears of joy. Rot! The inhabitants hang white rags on the roofs of their houses, tell us all the lies imaginable to prevent us obtaining water for our horses and let us plainly know we are aliens, and to keep our distance. Even the thousands of Turks we have captured, are fatter than we; they are excellently equipped, they have abundance of food and tobacco, their transports are overloaded with warm winter clothing. They are profusely supplied with ammunition for all arms, their air-service is excellent and has consistently been far superior to ours right up until Allenby came along and brought with him numerous and modern machines.

No, the papers have ignored our campaigns altogether until this great victory. Now they are full of skite and lies. Certainly we have passed occasional villages that have suffered so terribly from the Turks that no pen could describe their woe. But those villages and towns, so far, are in the minority. Also, the Turk is by no means the routed army. He is fighting us desperately, yard for yard, along a battlefront extending right across Palestine. We have particularly noticed the plentiful supply of bombs, aerial torpedoes, oils and petrols in the captured Turkish aerodromes. The great drums of petrol are branded "Vacuum Oil Company, America!"

Next morning—It rained last night, towards Jerusalem very heavily. That is great! There will be plenty of water in the wadi-beds for our horses. It makes us a bit miserable though; we have no tents, only a wee "bivvy" sheet to each man. A bivvy sheet is a strip of waterproof just long enough and wide enough for a man to spread on the ground so that he will not actually be "inside" wet mud. We tie four sheets together and make it a roof held up by our rifles.

…Greatest of boons—a tobacco issue has just arrived!

…We hear rumours that Hebron has fallen! What a triumphal advance for us! … Our Biblical expert has made the most of the "spell," wandering about the lines for miles picking up fragments of history. It appears that Tel-el-Safi, where the Turks have launched counter-attack upon counter-attack, is the site of the City of Gath, once famous as a splendour of the Philistines. In later centuries it was the Blance-garde of the Crusaders. We saw the place as precipitous whitish cliffs visible for many miles. It is a sort of Khyber Pass through the Judean hills to Jerusalem, a favourite haunt of peace-time robber bands.

We are much more interested in the line Nahr Rubin Wadi-el-Surar, where the Turkish battalions are advancing in desperate attacks. The centre of the line is Katra, with the village of El Mughar perched on its rocky hills nearby. The German engineers have boasted that when the English tackle Katra, they will find another Gibraltar. East of El Mughar is the junction of the Gaza and Beersheba railways with the main Nablus-Jerusalem railway. So we can easily understand that the Turks will fight like furies to keep open their fast closing roads of escape.

Rear view of the left flank outpost of the 5th Australian Light Horse Regiment on the heights of the clay hills in the Ghoraniye Bridgehead.

50

November 12th (I think) — Colonel Cameron with some brigade officers were watching from a hill, so we curious ones must go up too. We peeped through the glasses at the assault on Katra, miles away. By some miracle, the infantry had caught up to the mounted men away there too. We could see the infantry like school toddlers in long, very slow moving lines, advancing steadily to that grim hell that Katra must have been. Farther away, toy horsemen were galloping towards El Mughar. "Yeomanry!" said Colonel Cameron. We could not see where the Australians were operating. Over all was smoke and dust from which rumbled thunder.

About the 13th — Our brigade is ordered to ride across to the Judean hills and reinforce the Australian Mounted Division … We are sheltering behind low hills. The guns of the Turkish rear-guard are roaring close by … Yeomanry passed us yesterday, their horses nice and fresh. Our horses have bucked up wonderfully with couple of days' spell.

It has done us all a world of good too; we have had a chance to dress our septic sores. Eighty per cent of us are half rotten with them; but the flies won't get such a feed now.

Evening — The guns have quietened. There seems a breath of peace … The 7th occupied a village today in which the inhabitants crowded out and cheered lustily for the British.

Early morning, the day after — We moved off again shortly after dark and were soon riding past shadowy company. Infantry everywhere. Men in fire-lit groups were singing in that hearty fashion of the Tommies. As we rode amongst them our quick hoof-beats mingled with their laughter and their eager inquiries as they crowded around walking a few steps beside us. Nice chaps they were and we wished them good luck. We saw the secret of their phenomenal progress — lines and lines of motor-lorries! Through the night, some miles ahead, a fire started and spread in leaping sheets of flame as if a town were ablaze. It is still burning in columns of rolling black smoke clouding the tragedy of a retreating army. Everything this morning is very still — just here. We are moving out against the rear-guard at any moment. Bert is laughing in high glee. He has just had a little argument with the quartermaster and won. Bert is not as poddy as he used to be, but he's just the same old stick.

11 a.m. — We have travelled rapidly this morning; the Turks are so hard pressed that they cannot spare sufficient men to block us reinforcing troops. The prisoners we take now are haggard, their stubbly faces lined by exhaustion and despair. They have been fighting night and day for long past now while retreating—retreating—retreating. I get a chance to whip out the diary at occasional halts. We are passing very pretty little towns now, all red roofs and white walls with their pear hedge and orchards circling the little hill-tops. Fertile lands run between. Eagerly all hands watch the ground, and should a twig or bramble, a chip of wood be seen, there is an immediate rush of horses and men to secure the prize. We can't find enough wood to boil our quarts even. Water is still scarce. In the last convoy we captured were some German wagons, in one of which were cases of ointment. We tried it on our septic sores; the men crowding around that wagon reminded me of the flies crowding our sores. It is misery trying to sit the saddle. The ointment is grand. Within twelve hours most of the small sores grow scabs and soon after stop from running. The ointment burns like blazes.

The day after—We moved off again riding towards Junction station, the guns roaring the imminence of a counter-attack. With a quick preliminary attack then a whirlwind galloping charge we soon took a large village, the railway line running through it. Some of the Turks gamely tried to escape by running back along the line and the machine-gun bullets trilling off the rails made me imagine molten lightning skidding off greased iron. Here had been a fine German camp and 'drome. Lorries were still smouldering, one was the largest motor I have ever seen. Taubes lay there all twisted up and burnt. They left in frantic haste when they could not fly their 'planes away. Large quantities of machinery were blown up, huge stores of materials were blazing fiercely—for the camp had had "petrol to burn." A fine barracks room was stacked with swords … Guns are roaring; machine-gun fire rising in harsh waves. Shells come searching us out with a shriek and vicious crash almost as if they carry the hateful despair of the Turk … The 1st Light Horse Brigade has captured both Ramleh and Ludd … The Yeomanry Mounted Division and Camel Brigade have taken Abushusheh. The New Zealanders are in desperate fighting back on the coast towards Jaffa.

…Early this morning we rode on a pretty Turkish camp, the tents still standing, the little houses intact, the Black Crescent flying from a dozen buildings. We drove the Turks back, we rushed the place and they fought a running fight among the tents, dropping to their knees and snap shooting at us, doubling back to blaze at us in grey-clad groups around the corners of the buildings, running back again, tumbling over like shot wallabies, only surrendering when we were right on top of them, panting and glaring-eyed as they held up shaking arms. Their comrades are lining the hills just in front now, plastering us with shells. All arms of our army are pressing the broken Turkish armies without a moment's respite. A regiment of the 4th Light Horse Brigade has cut the Jaffa railway line and captured three trains; one was a hospital-train and in it was a man of the 8th Light Horse wounded in the back during the First Gaza battle.

A troop of us were sent on special duty miles away out to the right. It was intensely interesting riding just behind the fighting brigades, across but behind their line of fire. When we came to the foothills of Jillia near the Kezazah Pass, the scenery was rugged and grand. The Turks have fought most desperately to hold their mountain passes. Above the Wadi Surar, the hills overlook the Jerusalem road by Khuldah. Sheikh Musa Mountain towers over the road. By there, Turkish transports were in smashed confusion where portions of their army had retreated towards Jerusalem. Other portions retreating towards Ramleh, Ludd, and Jaffa have been smashed, but large forces have got away and still are determinedly fighting.

The men around here tell us that the Turks again deliberately fired on our Red Cross.

…We are back at brigade. A rumour is that we have to swing back again towards the coast and rejoin the Anzacs.

I think this is the 17th, Yesterday we rode by numbers of vividly pretty little red-roofed towns. The inhabitants are very fair skinned, mostly Jews. They are by far the most cleanly people we have yet met. They are very hospitable although they do charge us a hefty price for brown bread, honey, and tobacco. Lots of them have had a hard time from the Turks. They seem to live between two devils, the Turk and the Arab. Apparently the Turk prevents the Arab from massacring them outright, because the Jews are a very handy people to squeeze taxes from. Distantly we see the city roofs of "Jaffa the Beautiful"—very pretty in its hills and trees and orchards, even at this distance.

…We still ride on, to the muttering of the guns. We meet plenty of dead Turks and, increasingly now, wounded ones; sometimes forlorn little groups, haggard and dusty and inexpressibly weary. Badly hurt men are moaning, while others sit crouched or kneeling beside them; often a man holds up a shattered arm in mute entreaty while his comrades gaze at us wild-eyed as if actually afraid we might hurt them. Columns of prisoners pass us, captured at Ludd.

…Some of our boys got wine yesterday from the inhabitants of the little town nearby, and things are a bit lively. I think the town is Richon. Yesterday we rode through the place along a narrow road, the inhabitants in such queer garments lining the roads and shady lanes to stare at these brown, sleeveless soldiers. We must have seemed queer fighting men to them for they stared as if they had expected to see supermen, not rough-clad Australians. I don't think they could realize that we actually were the men who had driven back their taskmaster of centuries. They seem also to be on the verge of something they cannot believe, cannot understand: they tremble when they whisper of Jerusalem. It appears there is some prophecy, centuries old, that one day Jerusalem will fall and be taken from the Turk or from whatever infidel holds it.

The old neddies enjoy these rides through the shady lanes. We can tell by their willing movement, by the quick ring of their hooves that they think this land just fine.

The Turks have left numbers of these towns and their mild-looking inhabitants quite unharmed. All on account of taxes, we hear. But other places have suffered the dreadful penalties that always seem to tread upon the heels of a conqueror. Except when the British conquer — and they go to the other extreme.

We passed a quaint little building with "Hotel" prominently on a signboard, and other signs in Greek. A fair Hebe was leaning over the veranda with bared arms and a winning smile. She had won too, for the place was full of officers of the forces taking occupation, others seemed to be arriving per horse and lorry every minute.

These towns have plenty of flocks, plenty of wine, plenty of bread; the people are clean and civilized. We are coming into orchard towns now. The green oranges have given the whole regiment the tummy-ache. Rewinds me of the melons of the Bardawil.

A beautiful period of our ride was after crossing the Wadi Hanein. We rode through tall mimosa hedges, in perfumed bloom, into the colony of Nachalat (the heritage of Rueben). Crowds of white men, women, and children flocked the scented roads shouting: "Shallome!" "Shallome!" "Shallome!"

We are convinced that the Bible has gone to our enthusiast's head. He says the New Zealanders crossed the Nahr Sukereir by an old stone bridge built by the Crusaders. The Nahr Sukereir is a real river with flowing water, we don't believe it. There's one thing we are certain about though, the New Zealanders have been furiously opposed by the Jaffa Turks. All one day too at Ayun Kara they were fighting with the bayonet, charging and being charged. Ayun Kara is the place where Samson fought with the jawbone of an ass. I'll bet the En Zeds found the cold steel a deadlier weapon.

51

Stan arrived back from the mobile column yesterday with a fresh horse. He says that upon the Gaza redoubts are thousands of skeletons, relics of the Gaza battles, lying where they fell with their rusty equipment still upon them. Stan also says that the Turks played hell with the city, chopped down the trees in the parks and gardens and took all the rafters from the buildings to help strengthen the fortifications. What remains of the tank on Tank Redoubt is still there.

…Jaffa has fallen to the New Zealanders.

Next day—The Turk still fights grimly on. He is rather wonderful. We left our prettiest town yesterday and rode for miles through lanes densely hedged by prickly pear, on either side of which were orange orchards laden with golden fruit. Entering Ramleh, the road was chokingly dusty. It is a very large Arab town surrounded by luxuriant olive groves, its eastern streets are winding and narrow. The 1st Light Horse Brigade took this place a day or two ago. Dead Turks are lying about, so dusty now by the constant movement of horsemen as to lie unrecognisable but as dusty logs. The Arabs are too lazy to bury them even for their own health's sake and the live Turks are keeping our own chaps too busy for anything else but fighting, for the time being. The only imposing buildings in the place are the Greek churches and the old-time Tower of the Forty Martyrs. Crumbling tombs around one old ruined church look as old as the Judean hills. The modern church had a white flag drooping from the cross. I hope no Australian church will ever have to hang such a symbol of disgrace and despair.

Ludd is a rather similar town to Ramleh. Both places are rich in memories of the Crusaders. Ascalon and Ekron and Esdud and other old-time terrors are round about. What with Samson, and the Israelites and Philistines, Abraham and Moses and Isaac and the Australians and New Zealanders, not to mention the Yeomanry and Coeur de Lion, a man doesn't know in whose dust he is riding. These places were crowded with troops, and the dirty inhabitants scowled at us plainly wishing they were men enough to cut our throats.

Farther on we rode through more olive groves, the growl of guns ever beckoning just ahead … We are ready to move off again It rained last night. The farmers amongst us are very interested in this country's cultivation.

About 1p.m.—We moved off, riding through flat, cultivated lands, and presently passed the white city of Jaffa on our left. To our right are the Judean hills, now bearing in towards the coast. The smaller foothills are dotted with villages, many of them barely a stone's throw from one another, all surrounded by fruit-trees. The Turkish guns are shelling us…

…An hour ago we were transported with delight. We rode down a beautiful avenue of Australian gums, with fine two-storied farm houses on either side, big stacks of hay everywhere, gardens. fowls, and the atmosphere of a perfect farming prosperity over all. We have camped here for dinner and are not at all surprised at the fine strapping girls. Where a healthy gum-tree can grow it is a certainty a good-looking girl will spring up. But they are Germans! Close by is a schoolhouse; its bell persists in clanging every quarter of an hour. Such a barefaced way of telling the German gunners when our troops are coming. We hear laughing school kids' voices. Dotted all over the place are snipers' cunningly concealed possies, each with its tell-tale heap of empty cartridge shells. At a house corner is a pile of shells where a machine-gun was turned on our chaps as they advanced.

Next day—Our Bible enthusiast wangled a trip to Jaffa. The inhabitants had a dreadful time under Jemal Pasha. He cast out forty thousand of the population, only allowing ten thousand red-hot Turkish sympathizers to stay in the city. Jaffa is very beautiful, set in the hills by the coast, surrounded by its gardens and fruit-groves. By the old-time city is a stone quay, as good as when it was built, to tie up Crusaders' ships. Just off-shore is the rock of Andromeda, where that young lady was tied while the tide came up. In sight of this rock is Simon the Tanner's house. Napoleon was caught in a tight corner at Jaffa, so he massacred his prisoners and poisoned his sick.

When the En Zeds took the city there were some thousands of poor little devils of orphans there, made so by the Turks. Hundreds died of starvation. We were much surprised when the Bible enthusiast told us that Miss McConachy, the Red Cross worker we all knew at the Soldiers' Club in Esbekiah Gardens, Cairo, was there in charge of some mission trying to save the orphans. The first thing the New Zealanders did when the town surrendered was to find her house and place a guard over it.

…The 52nd Infantry Division attacked Katra. The Yeomanry Mounted Division, covered by the Camel Brigade, crossed the Nahr Rubin near Yebna where one brigade, the Bucks, Berks, and Dorsets swung to the east and hell for leather charged El Mughar in the face of gun, machine-gun, and rifle fire. We hear they broke the Turkish right and centre in superb style. Good luck to them.

Next day (about 10 a.m.)—In this German colony, all the young men have long since enlisted, some fighting for the Fatherland in France, others in Palestine. We had taken some of these very men prisoners at Beersheba. The womenfolk were eager to learn of the Palestine fighting. Scornfully and fearlessly they assured us that we will soon be driven back past Beersheba, out into the desert again, where the jackals will fight the vultures for our bones. However, for the time being they eagerly sell us brown bread, butter, honey, milk, fowls, sucking-pigs—and hay for our horses. We wonder what the Belgian civilians would have thought had they been treated similarly.

We keep our horses well up against the houses, in amongst the gum-trees. It is good for our eyes and strange to see the brown groups of contented Australian horses shading amongst these healthy gums in this clean place.

Afternoon—Came a heavy splintering—crash! High explosive and earth spouted from just in front of the houses nearest the Turkish positions. Now comes another, another, another! Two of our signallers are down.

When our fellows group just a wee little bit away from any of the houses a big shell most uncannily finds them out. We know of course that in the colony there are cellars with underground telephones connected with the Turkish lines. The women do the above-ground spying, and Germans dressed as women. Also their Intelligence men can peep at us from the windows. The English would never allow their troops to enter houses. No wonder they call us fools.

Late afternoon.—I've had the worst shock since I enlisted. Because I'm sick, I suppose. I half wake up at nights with the doctor kneeling beside me, one arm holding up my neck while the other pokes a dashed long finger down my throat. Then he crawls away to fill other poor devils with quinine pills.

Anyway, from a little village to our right, came a sudden burst of machine-gun fire and away from the village at full gallop came a troop of our men. This machine-gun and rifle fire at such close range surprised us, it was almost behind us too. We pressed closer in amongst the friendly gums. They machine-gunned us for an hour, searching the trees with exact range. Then came a pattering all around our troop and hundreds of little somethings hissed past my ear and buried themselves by the horses' hooves. We sprang up immediately, but for some unexplainable reason the machine-gunning switched on to another troop. I think that the gunners, having to shoot between the trees, could not see the results of their shooting, and by the time they were phoned each target had very quickly moved elsewhere. This machine-gunning into our backs by men in well-hidden cellars was awful. I think we all felt in a nasty state before the welcome dark came. I was scared just sick.

…With darkness came rain. We filed out and formed an outpost line in front of the colony, expecting a Turkish counter-attack all through the night. They must have known to a man our strength, one regiment holding that long, thin line.

With the darkness I felt all right again. It was a fairly miserable night, dark with ceaseless rain that hissed in whispering noises; a man's neck became kinked through listening for stealthy footsteps while he lay on the wet ploughed ground. I crawled all over the place, but I'm blessed if I could locate an even piece of ground to lie on anywhere. About twelve o'clock a flare blazed up in front. We peered, gripping our rifles, expecting the attack. Had we only known it, it was the Turkish rear-guard signal to retire.

An Australian Light Horse regiment near Jerusalem.

52

Dawn broke, black and drizzling. The regiment posted its outposts—a nasty job theirs, crouched in the mud staring out into a bitter dawn almost as black as night—then returned to the colony, where we boiled up. The section bought some appetizing brown bread from a German family whose most treasured possession was a picture of the Kaiser over the mantelpiece. We bought honey too, and butter. What a great breakfast that was! Bert went scouting and returned whistling with a billycan full of milk which made some warming cocoa, with such a gorgeous influence on Stan, that he combed his nice, curly hair. Then the regiment lit its pipe and rode out to where rifle-shots told that the patrols had located the Turks. From a hill we got a silver-grey view of the country and a misty battle raging to the right towards Jerusalem. To our left was the sea, and Jaffa with its towns and villages now peeping from rain squalls. The shells the Turks sent over exploded all squaggy in the mud. Johnson had his face plastered like a nigger; the force nearly drove his eyes into the back of his head. Mud is soft, but he lost four teeth and half of one lip. The germs in the mud do not help our wounded chaps any, either.

Later on, the 7th Regiment relieved us. We gladly left them to the shells and machine-guns; gladly enough, too, we left Wilhelmina and retired on Ramleh for the remainder of the day.

Next day—It rained heavily all yesterday and last night. Bitterly cold. But this morning is glorious. Under the wagons lie huddled in the mud some Egyptian Labour Corps men who have perished in the night. We are hardier by far than the native Egyptians.

We are out again after the Turkish rear-guard and expect to be within rifle-shot at any moment. We are well supplied with tucker now. The transport organization and the A.M.C. during these operations has been splendid. We all feel proud of Allenby.

2 p.m.—We are lining the slopes near Jaffa, our guns blazing at the Turk. His trenches are in the orange orchards. His bullets are whistling, whistling, whistling.

Since the attack on Beersheba, the Desert Mounted Corps has driven the enemy back a hundred and twenty miles, captured thousands of prisoners, numerous convoys, and tremendous war-supplies; taken seventy guns and fought five big battles, not to mention the every day and night fighting. So the corps have their tails up.

Throughout the area behind us are flocking thousands of

townspeople, back to their homes in Jaffa. They have been hiding for over two years past from the Turk, in the villages of the plains, amongst the Judean hills, anywhere to which they could escape. The lucky families own a scraggy camel, or patient horse, or poor little donkey, quite hidden under a load of family treasures. It is comically pathetic to watch a skinny camel lurching along with a wardrobe swaying on each side of him, another comes rolling like a ship at sea with a sofa on one side, all the chairs, pots, pans, and mattresses of a family on the other. Men, women, and children in flocks, trudging wearily along the dusty roads. Some of the girls, in their threadbare tribulation, look pathetically nice.

…Arab hawkers are coming out even to the firing-line to sell us Jaffa oranges, bread, honey, sweetmeats, and all sorts of stuff. They charge 3d. for a box of matches. Shells are screeching across very thickly now. They are terrifying damn things when a man senses one about to drop in his vicinity.

14th Australian General Hospital, Cairo.

December 23rd—It's a long time since I've made an entry in the old diary. Here goes for the last. The army was holding the line right across Palestine from Jaffa to Jericho right up near Jerusalem. Our brigade was in the line outside of Jaffa. I felt wretched through some form of malarial fever—Jaffa fever they call it. The last phase of activity I remember distinctly was watching the New Zealanders charge through the Turkish shells and take a ridge towards our left. I can remember the bayonets flashing, the hair-raising New Zealand war-cries, and the screaming "Allahs!" of the Turks. Next morning the Turks rushed up fresh reinforcements and counterattacked under a fearful barrage. The New Zealanders were literally flung from the ridge, but we had the pleasure of watching our own artillery mow down the massed Turks. Their remnants, though being constantly reinforced, dived into the trenches for shelter from the hail of iron. Fronting us, the country was dotted with villages, one very pretty Jewish one, Muelebis, being all smoke from bursting shells. Groves of dark green gums were everywhere. The inhabitants are very jealous of each individual tree. They import the Australian "sucker" gum, the gum that our pastoralists find almost impossible to kill by ring-barking. When these people cut down one tree, six young ones spring up from the old stump. When I was "suckering" in Australia I little dreamt that one day I would be in a strange land where people would treasure as more precious than gold, the very trees that we sought so hard to kill as pests.

From then on I was too sick to notice or care what was happening. The Turks were reinforced again and again. I've got a dim idea that fresh troops were pouring in from the Balkan and Russian fronts. We dug ourselves in. Shells rained day and night, all night long the whole line roared under rifle and machine-gun fire heightened into an inferno under attacks by hand grenades. Our own artillery right up amongst us nearly burst my ears. If hell is any worse then I don't want to go there. The blasted 'planes added their racket to the frightening uncertainty of life. The doctor, with a lot of trouble, managed to pull me through and one morning, feeling much better, I went on sick parade to get the wretched septic sores dressed.

I was sitting down waiting my turn, talking to the Red Cross sergeant on the gully-bank when—*whee-eee-eezz crash!* Instinctively we had thrown ourselves on our faces but the shell exploded too close. I spent awful seconds wondering whether I was hit mortally. A whirlwind of bells was ringing within my head, and I knew that until the bells quietened I would not be able to think clearly at all. The first recognizable feeling was the numbness, with my mind trying to telephone to all parts of the body to find out which were still there and which broken. Then quick as anything I realized I had a good fighting chance though I still felt all numb down my back. The bloody hole in the arm and thigh did not matter. Soon I was swathed in bandages. By the rules of the game I should have been blown to pieces but there was not even a bone broken, just a dozen shell splinters.

They put me into a sand-cart ambulance and soon I lost some illusions, having all my time taken up in desperate snatches at the cart to ease the bumps when going over the slightest uneven ground. At nightfall they left me at a dressing-station, gave me a good feed, probed out some stray shell splinters, and fixed the bandages the cart had jolted loose. Then came a motor ambulance and the loss of more illusions. Motor ambulances look easy things as they glide over the roads, but for a wounded man, at the rate the drivers drove their long string of cars, every yard was torture. I hung on to the car and prayed for the awful journey to end. We pulled up at the Australian Casualty Clearing Station, I think, in Jaffa. Some doctors had a look at us, stretcher-bearers carried us into a huge bare hall, put the stretchers down, gave us a meal and left us alone. The hall was enormous, a great dome-shaped place with a glass chandelier sparkling away up in the roof. The floor was of cold marble. The great bare room was semi-dark with mysterious, lighted corridors branching out from it. Hum of talk came from these corridors, they were occupied by men convalescing from malaria.

It was a strange night lying there in a sort of restful fever, guessing the war was over for me; strange too after so many nights camping out under the stars.

Next day a convoy of us set out for Ramleh. I don't know any particulars, just hung on to the stretcher to try and ease the bumps. I managed to roll over on my belly and had a better chance, being thus able to grip the stretcher with both hands. I pitied the poor devils who were too much hurt to protect themselves; but my heart went out to the poor fellows with fractured limbs. We were all exhausted long before the convoy pulled up by some old eastern building. Here, with great labour, we were carried up a spiral flight of steps; they seemed to wind around time inside of some big old castle tower. It opened out at last into a narrow, musty stone passageway. The only light came from dim loopholes high up in the cold grey walls. I remember the bearers' boots shuffling on the stone flags. The corridor opened out into dim little cubicles of bare stone, having tables over which white-gowned doctors were bending. There was a captured Turkish doctor here and he was attending some wounded Turks. Others moaned in the shadows. There was a young Assyrian girl too, badly hit in the back by shrapnel. A grey-bearded old village doctor crouched by her, but all he seemed to do was attend to her moaning wants with water from an earthen jar. Then he would crouch, his head bowed, his long beard pressed upon his gown, his lips trembling as if whispering to something unseen. The girl's poor eyes, tear-dimmed with the fear of death, gazed at me as if she thought that surely God must send someone to help her. Two Tommy doctors bent over her. I knew by the expression on their faces—even by the gentleness of their fingers— that every possible thing that could be done by man, would be done to save her. Then the Tommies went to a lot of trouble over me. It was through their careful dressings, I am certain, that the wounds did not go septic during the journeys ahead. They even cleansed the septic sores.

About midday, a long convoy started for a distant dressing-station. Soon some of the fellows began to cry with the pain of the jolting, but nothing, short of a machine-gun, would have made the drivers drive slower. Perhaps it was because we had long distances to go and must get there, but each drive was hell. At dusk we arrived at a big field hospital, thanking God for the finish of another awful journey.

We were placed in rows in big tents, the worst cases had their wounds dressed. They gave us bully-beef and bread and a mug of tea, but I was too weak to care. Next day we travelled on to another big clearing-station but the English doctor here saw to it himself that the orderlies gave us a decent meal. He then went away and came back loaded with oranges. A padre came too and handed a packet of cigarettes to each man. We blessed those two men.

I forget whether it was a few hours later or next morning, but we were in the motor ambulance again, being driven through hell. I think that was the last journey: I was too far gone to know. One evening we pulled up at a big English clearing-station and were told that the rest of our journey would be by train. We could have cried with relief. We were carried into a Red Crescent train, captured from the Turks. On each side of the train were wooden bunks like bunks in a close-packed ship. The wounded girl was placed in the bunk opposite me. She moaned sometimes in such a heartbroken way, her eyes had a great fear. I felt so sorry; all us hurt chaps felt so terribly helpless. We knew that she must die.

At last we moved off. There was no jolting. I lay on my right side, and closed my eyes in intense relief. An orderly came around with hot cocoa, bread and butter. It was great. Afterwards I lay quite still though the cursed lice were biting.

We sped through the desert all night—how vividly I remember that desert! I shall never forget!

Next morning we pulled up at Kantara on the Canal. Under what differing circumstances I have seen that huge Base Camp! By motor ambulance, we were taken to a clearing-station. But these drivers drove slowly! At the clearing-station we were well looked after. Most of the wounded went off that night to Cairo in the Hospital Train, but a number of us were kept back twenty-four hours to pick up a little strength. Then ambulances again across the pontoon bridge to the railway station. How comfortable that big Red Crescent train was. Broad bunks actually with white sheets on them, orderlies dressed in white, even nurses! A real good meal and cigarettes! My leg started a haemorrhage. A Tommy doctor stood by me through the night so that I would not tear the bandages off. He was such a decent, patient chap, and I gave him a hell of a time. The last miles of that journey were hell to me, every second.

Then motor ambulance again through the streets of Cairo, and finally here.

It is Christmas Eve. The nurses, the doctors, the Red Cross and the civilized amongst the citizens of Cairo, cannot do enough for us.

Port Said Hospital.

January 2nd, 1918—I am to be returned to Australia as unfit for further service. Thank heaven!

ION L. IDRIESS.

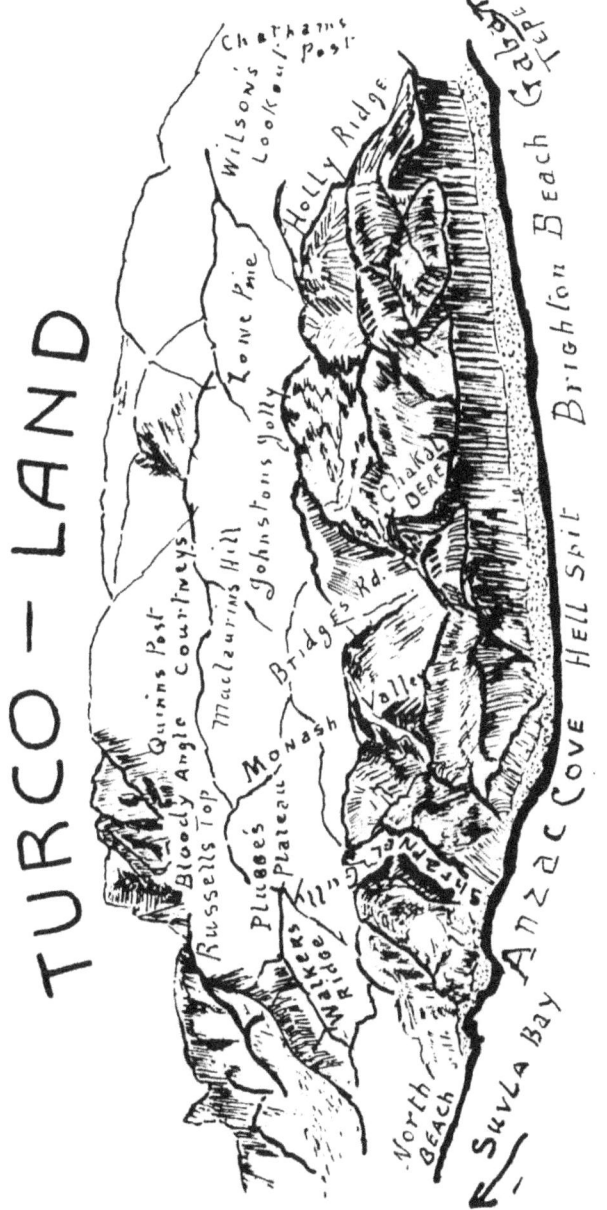

The author's impression of Gallipoli.

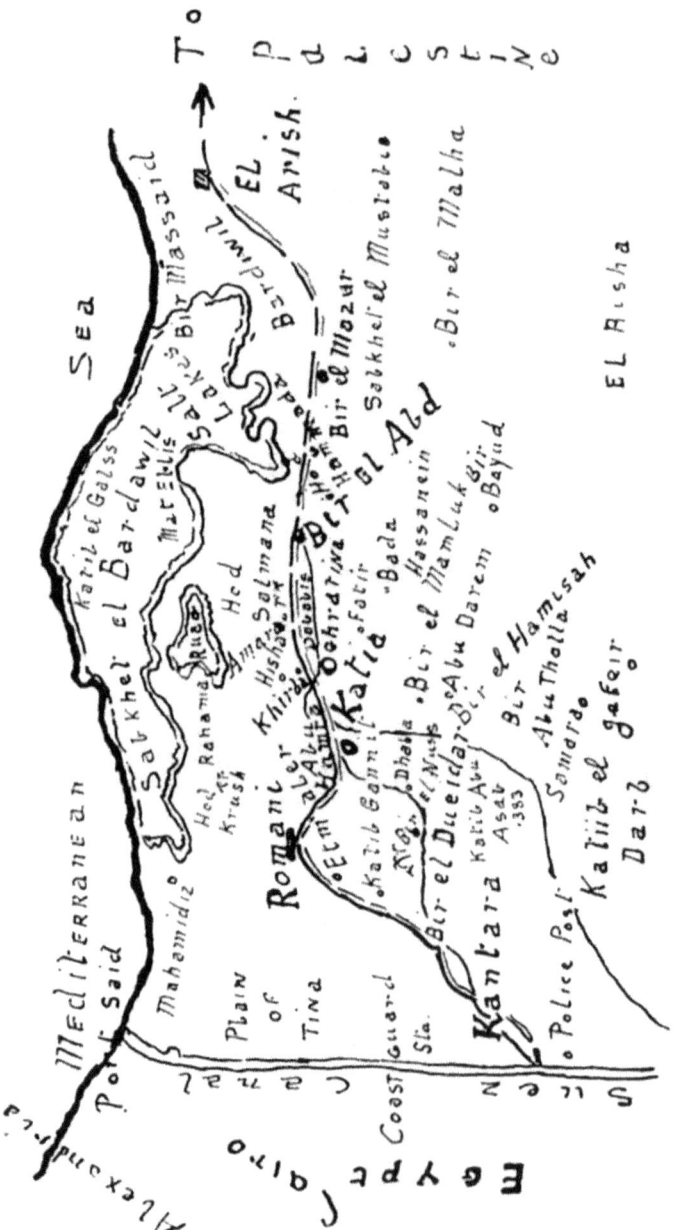

The author's sketch map of the Sinai desert.

www.ingramcontent.com/pod-product-compliance
Lightning Source LLC
Chambersburg PA
CBHW032038150426
43194CB00006B/330